CHEMOINFORMATICS
Advanced Control & Computational Techniques

CHEMOINFORMATICS
Advanced Control & Computational Techniques

Hossein G. Gilani, PhD, Katia G. Samper
and Reza K. Haghi

Apple Academic Press

TORONTO   NEW JERSEY

© 2013 by
Apple Academic Press Inc.
3333 Mistwell Crescent
Oakville, ON L6L 0A2
Canada

Apple Academic Press Inc.
1613 Beaver Dam Road, Suite # 104
Point Pleasant, NJ 08742
USA

Exclusive worldwide distribution by CRC Press, a Taylor & Francis Group

International Standard Book Number: 978-1-926895-23-9 (Hardback)

Printed in the United States of America on acid-free paper

Library of Congress Control Number: 2012935657

Library and Archives Canada Cataloguing in Publication

>Chemoinformatics: advanced control & computational techniques/edited by Hossein G. Gilani, Katia G. Samper and Reza K. Haghi.
>
>Includes bibliographical references and index.
>ISBN 978-1-926895-23-9
>1. Cheminformatics. I. Gilani, Hossein G II. Samper, Katia G III. Haghi, Reza K
>
>QD39.3.E46C44 2012   542›.85   C2011-908703-0

Trademark Notice: Registered trademark of products or corporate names are used only for explanation and identification without intent to infringe.

This book contains information obtained from authentic and highly regarded sources. Reprinted material is quoted with permission and sources are indicated. A wide variety of references are listed. Reasonable efforts have been made to publish reliable data and information, but the authors, editors, and the publisher cannot assume responsibility for the validity of all materials or the consequences of their use. The authors, editors, and the publisher have attempted to trace the copyright holders of all material reproduced in this publication and apologize to copyright holders if permission to publish in this form has not been obtained. If any copyright material has not been acknowledged, please write and let us know so we may rectify in any future reprint.

All rights reserved. No part of this work covered by the copyright hereon may be reproduced or used in any form or by any means—graphic, electronic, or mechanical, including photocopying, recording, taping, or information storage and retrieval systems—without the written permission of the publisher.

Apple Academic Press also publishes its books in a variety of electronic formats. Some content that appears in print may not be available in electronic format. For information about Apple Academic Press products, visit our website at www.appleacademicpress.com

Contents

List of Abbreviations vii
Preface ix

1. A Thermodynamic Model for Vapor–Liquid Equilibrium Data for the Two Binary Systems 1
2. Practical Hints on Measurement of Densities and Excess Molar Volumes 7
3. A Mathematical Model to Control the Liquid–Liquid Equilibrium Data 11
4. Computer Calculations for Multicomponent Vapor–Liquid and Liquid–Liquid Equilibria 27
5. Densities and Refractive Indices of the Binary Systems 43
6. Potential Applications of Artificial Neural Networks to Thermodynamics 49
7. A Note on Application of Non-random Two-liquid (NRTL) Model 61
8. Some Practical Hints on Application of UNIQUAC Solution Model 65
9. Control of Liquid Membrane Separation Process 75
10. Development of Artificial Neural Network (ANN) Model for Estimation of Vapor Liquid Equilibrium (VLE) Data 91
11. Some Aspects of a Fluid Phase Equilibria and UNIFAC Model 101
12. Molecular Thermodynamics Process Control in Fluid-phase Equilibria 115
13. Estimation of Liquid-liquid Equilibrium Using Artificial Neural Networks ... 125
14. Optimization of Process Control of Water + Propanoic Acid+ 1-Octanol System 135
15. Practical Hints on Optimization of UNIQUAC Interaction Parameters 147
16. A Mathematical Approach to Control the Water Content of Sour Gas 155
17. Modling and Control of Thermodynamic Properties by Artificial Neural Networka (ANNs) 165
18. A Study on Liquid–Liquid Equilibria using HYSYS and UNIQUAC Models ... 173
19. Optimization and Control of Laboratory Production of Ethanol 185

   Index 199

List of Abbreviation

(ANOVA)	Analysis of variance
(ANNs)	Artificial neural networks
(BOD)	Biochemical oxygen demand
(B.M)	Bukacek–Maddox
(CCD)	Central composite design
(CCWP)	Cheese whey permeate
(COD)	Chemical oxygen demand
(DCW)	Dry cell weight
(EOS)	Equations of state
(2EH)	2-Ethyl-1-hexanol
(FNNs)	Feed-forward neural networks
(GC)	Gas chromatography
(GMDH)	Group method of data handling
(H–C)	Hayden and O'Connell(HFEs) Hydrofluoroethers
(IROST)	Iranian Research Organization for Science and Technology
(LLE)	Liquid–liquid equilibria
(M–S)	Marek and Standart
(MAD)	Mean absolute deviation
(MAE)	Mean absolute error
(MSE)	Mean square error
(MTBE)	Methyl *tert*-butyl ether
(MLP)	Multi-layer perceptron
(NTC)	Na-thioglycolate concentration
(NBA)	n-Butanol
(NNs)	Neural networks
(NRTL)	Non random two-liquid
(OF)	Objective function
(PNNs)	Probabilistic neural networks
(RBF)	Radial basis function
(RF)	Response function
(RMSE)	Root mean square error
(TAOH)	Tert-amyl alcohol
(TAME)	Tert-amyl methyl ether
(TBA)	*tert*-butanol

(TCD)   Thermal conductivity detector
(UNIQUAC) Universal quasi-chemical
(VLE)   Vapor–liquid equilibrium

Preface

This book provides a broad understanding of the main computational techniques used for processing chemical and biological structural data. The theoretical background to a number of techniques is introduced and general data analysis techniques and examining the application of techniques in an industrial setting, including current practices and current research considered. The book also provides practical experience of commercially available systems and includes a small-scale Chemoinformatics related projects.

The book offers scope for academics, researchers, and engineering professionals to present their research and development works that have potential for applications in several disciplines of chemoinformatics and science. Contributions ranged from new methods to novel applications of existing methods to gain understanding of the material and/or structural behavior of new and advanced systems.

This book will provide innovative chapters on the growth of educational, scientific, and industrial research activities among chemical engineers and provides a medium for mutual communication between international academia and the industry. This book publishes significant research reporting new methodologies and important applications in the fields of chemical informatics as well as includes the latest coverage of chemical databases and the development of new computational methods and efficient algorithms for chemical software and chemical engineering.

1 A Thermodynamic Model for Vapor–Liquid Equilibrium Data for the Two Binary Systems

CONTENTS

1.1 Introduction 1
1.2 Thermodynamic Model 2
1.3 Results and Discussion 4
1.4 Conclusion 4
Keywords 5
References 5

NOMENCLATURES

T	Temperature (°C)
a_{ij}	Wilson parameteres (J/mol)
K_A	Dimmerization constant
P	Total pressure (kPa)
x_i	Liquid phase composition (–)
y_i	Vapor phase composition (–)
V_i	Molar volume (ml/mol)
z_i	Correction factor (–)

Greek letters

γ_i	Activity coefficient (–)
μ_i	Chemical potential (J/mol)
φ_i	Fugacity coefficient of component i (–)

1.1 INTRODUCTION

In the past decades, hybrid processes combining reaction and separation mechanisms into a single integrated operation became increasingly interesting for designers in chemical industry. Such combined processes are usually called reactive separation processes. The etherification reaction is a typical example of such a process which uses carboxylic acids like formic or acetic acid as raw materials. For reliable design of a separation unit, accurate experimental data on vapor-liquid equilibrium (VLE) are required. In many practical systems, the interactions between the molecules are quite strong due to charge transfer and hydrogen bonding. This occurs in pure components such as alcohol, carboxylic acids, water and HF and leads to quite different behavior of vapors of these substances. Considering the interactions so strong that new chemical species are formed, the thermodynamic treatment assumes that the properties deviate from an ideal gas mainly due to the associating species. These interactions may strongly affect the thermodynamic properties of the fluids. Thus, the chemical equilibria between clusters should be taken into account in order to develop a reliable thermodynamic model.

1.2 THERMODYNAMIC MODEL

For phase equilibrium at constant pressure and temperature, the chemical potential of each component must be equal in both phases, namely:

$$\mu_i^v = \mu_i^l, \quad i = 1,\ldots,n \tag{1}$$

Equation (1) can be rewritten in terms of measurable variables as:

$$y_i \varphi_i P = x_i \gamma_i P_i^{sat} \tag{2}$$

where the activity coefficients may be represented by a suitable model and the fugacity coefficients φ_i can be calculated from an equation of state:

$$\ln \varphi_i = \frac{1}{RT} \int_0^P \left[\left(\frac{\partial V}{\partial n_i} \right)_{T,P,n_{j \neq i}} - \frac{RT}{P} \right] \tag{3}$$

For this purpose, the virial equation of state is frequently used at low and medium pressures. However, for systems containing polar compounds with strong molecular association in the vapor phase such as acetic acid, the association effect must not be neglected in VLE computation. In a general method for prediction of the second virial coefficients, Hayden and O'Connell (1975) (H–C) takes into account these effects by using a chemical theory. In this study the approach of Marek and Standart (M–S) was used. Marek and Standart (1954) handled the problem by treating the association of molecules as a chemical reaction. Considering the dimerization of acetic acid (or formic acid) in the vapor phase and assuming otherwise perfect gas phase behavior, they introduced a correction factor (Z_i):

$$y_i Z_i P = x_i \gamma_i P_i^{sat} \tag{4}$$

where

$$Z_A = \frac{1 + \sqrt{1 + 4K_A P_A^{sat}}}{1 + \sqrt{1 + 4KPy_A (2 - y_A)}} \tag{5}$$

for an associating component A, and:

$$Z_N = \frac{2\left[1 - y_A + \sqrt{1 + 4KPy_A(2 - y_A)}\right]}{(2 - y_A)\left[1 + \sqrt{1 + 4KPy_A(2 - y_A)}\right]} \quad (6)$$

for a non-associating component N.

In equations (5) and (6) K_A is the dimerization equilibrium constant of pure component A, and K is the dimerization equilibrium constant of component A in the mixture. As discussed by Marek (Marek and Standart, 1954), it is a good assumption to set $K = K_A$. Dimerization constant for acetic acid obtained from Venimadhavan et al. (1999):

$$\log K = -12.5454 + 3166/(273.15 + T) \quad (7)$$

Dimerization constant for formic acid obtained from Bessling et al. (1997):

$$\ln K = -18.22 + 7099/(273.15 + T) \quad (8)$$

The following Wilson model was selected for liquid phase non-ideality:

$$\ln \gamma_1 = -\ln(x_1 + x_2\Lambda_{12}) + x_2\left(\frac{\Lambda_{12}}{x_1 + x_2\Lambda_{12}} - \frac{\Lambda_{21}}{x_2 + x_1\Lambda_{21}}\right) \quad (9)$$

$$\ln \gamma_2 = -\ln(x_2 + x_1\Lambda_{21}) - x_1\left(\frac{\Lambda_{12}}{x_1 + x_2\Lambda_{12}} - \frac{\Lambda_{21}}{x_2 + x_1\Lambda_{21}}\right) \quad (10)$$

$$\Lambda_{ij} = \frac{V_i}{V_j}\exp{-\frac{a_{ij}}{R(273.15+T)}} \quad (i \neq j) \quad (11)$$

The molar volumes V_i (ml/mol) of acid acetic, acid formic and water which are calculated by Racket equation are shown in Table 1.1.

TABLE 1.1 Molar volumes Vi (ml/mol) of acetic acid, formic acid, and water.

	$V_i (ml/mol)$
Acetic Acid	57.54
Water	18.07
Formic Acid	37.91

The regression was performed by means of the minimizing difference between experimental and calculated data in the objective function:

$$O.F. = \frac{1}{NP}\sum_i\left[(y_i^{cal.} - y_i^{exp})^2 + \frac{(T_i^{cal.} - T_i^{exp})^2}{T_i^{exp}}\right] \quad (12)$$

1.3 RESULTS AND DISCUSSION

All the experimental binary values and calculated deviations in liquid and vapor compositions and temperature, (calculated using the equation 4), are summarized in Table 1.2 where also the mean absolute deviations in the individual variables are reported. The resulting Wilson parameters optimized in this work for the two binaries are listed in Table 1.3.

TABLE 1.2 The experimental and literature data of formic acid (1) and water (2) mixture at 101.32 kPa.

P(Bar)	x_1^{exp}	y_1^{exp}
0.123525	0.0000	0
0.126798	0.0667	0.090812
0.130095	0.1333	0.177044
0.133376	0.2000	0.259034
0.136656	0.2667	0.337087
0.162898	0.8000	0.411481
0.166179	0.8667	0.482466
0.169459	0.9333	0.848334
0.172723	1.0000	0.900885

TABLE 1.3 The Wilson parameters for binary systems of acetic acid (1) + water (2) and formic acid (1) + water (2) system.

	a_{12}	a_{21}
acetic acid (1) + water (2)	−180.83700	813.09710
formic acid (1) + water (2)	1180.8040	−310.1060

Vapor phase imperfection and association of acetic acid and formic acid were both taken into account during this evaluation by employing the M–S method. Taking into account dimerization of acids in the gas phase gives correct VLE behavior. In particular, comparing with the vapor-liquid results from Alpert and Elving (1949) and Ito and Yoshida (1963) there is no azeotrope and less y_i values than x_i. Nevertheless, comparing the model predictions with the vapor-liquid data of Ito and Yoshida (1963) for formic acid-water system, it accurately describes occurrence of an azeotrope.

1.4 CONCLUSION

The associating systems of acetic acid with water and formic acid with water at constant pressure are considered in this chapter. Strong interaction between acids in gas and dimer formation are also included in the phase equilibria prediction. Gas phase non-ideality was taken into account by M–S method and Wilson equation was used for liquid phase non-ideality. Mean absolute deviations between experimental data and the data correlated using the Wilson equation are, in overall averages, 0.071 mol fraction in liquid phase and 0.0158 mol fraction in vapor phase; and the mean absolute deviations of 0.057 K were found for temperature. The model correctly describes behavior of binary systems specially formic acid-water system which shows an azeotrope

KEYWORDS

- Binary systems
- Esterification reaction
- Marek and Standart method
- Thermodynamic model
- Vapor-liquid equilibrium

REFERENCES

Alpert, N. and Elving, P. J. (1949). Vapor-liquid Equilibria in Binary Systems. Ethylene Dichloride-Toluene and Formic Acid—Acetic Acid. *Ind. Eng. Chem.* **41**, 2864–2869.

Bessling, B., Schembecker, G., and Simmrock, K. H. (1997). Design of Processes with Reactive Distillation Line Diagrams. *Ind. Eng. Chem. Res.* **36**, 3032–3042.

Hayden, J. G. and O'Connell, J. P. (1975). A Generalized Method for Predicting Second Virial Coefficients. *Ind. Eng. Chem. Process Des. Dev.* **14**, 209–216.

Ito, T. and Yoshida, F. (1963). Vapor-liquid Equilibria of Water-Lower Fatty Acid Systems: Water-Formic Acid, Water Acetic Acid and Water-propionic Acid. *J. Chem. Eng. Data* **8**, 315–321.

Marek, J. and Standart, G. (1954). Vapor-liquid Equilibria in Mixtures Containing and Associating Substance. Equilibrium Relationships for Systems with an Associating Component. *Coll. Czech. Chem. Commun.* **19**, 1074–1084.

Venimadhavan, A., Malone, C., and Doherty, J. (1999). Bifurcation study of kinetic effects in reactive distillation. *AIChE J.* **45**, 546–556.

2 Practical Hints on Measurement of Densities and Excess Molar Volumes

CONTENTS

2.1 Introduction 7
2.2 Experimental...7
2.3 Results 8
2.4 Conclusion..10
Keywords 10
References..10

2.1 INTRODUCTION

Excess molar volumes of V^E of 1,2-ethanediol + water were measured at $p = 0.1$ MPa with a faithful copy of the vibrating tube densimeter DMA 602 from Anton Paar. All binary mixtures were measured at the temperatures (308.2, 313.2, and 318.2) K. Values of V^E are negative for all the mixtures studied over the whole concentration range and for all temperatures. Results were correlated by polynomial equations of Redlich and Kister (1948).

In this chapter, densities and molar excess volumes of binary (1,2-ethanediol + water) systems in the temperature range of 308.2–313.2 K at atmospheric pressure are reported. The measurements are part of a long-term study to examine the dependence of excess volumes on temperature. In the earlier papers (Dean, 1985; Douheret et al., 1991; Geyer et al., in press; Kapadi et al., 2000; Nitta et al., 1977; Redlich and Kister, 1948; Span and Wagner, in press) extensive database on the thermodynamic interactions in binary mixtures of 1,2-ethanediol with water was presented. The aim of this work is to provide a set of values for the characterization of the molecular interactions of these mixtures and to examine the effect of the size of alkyl group of the diol.

2.2 EXPERIMENTAL

The 2,3-butanediol was supplied by Merck (analytical grade, mass fraction purity 0.99). Water, used as one substance to calibrate the densimeter was filtered four times (conductance < 1 ms). n-Hexane was supplied by SDS (analytical grade, mass fraction purity 0.992) and was used as the second substance for calibration. All materials used for measurement were used without further purification. All liquids were first degassed ultrasonically and additionally under reduced pressure before an experiment. Densities ρ of pure components and mixtures were determined using a vibrating tube densimeter, a faithful copy of model DMA 602 of Anton Paar. The measuring cell consists of original components, which are not made of glass.

2.3 RESULTS

The experimental values of the densities and the excess molar volumes V^E of the binary mixtures are given in Tables 2.1. The quantity of V^E for a mixture of two components was calculated from:

$$V^E = \frac{x_1 M_1 + x_2 M_2}{\rho} - (x_1 V_1 + x_2 V_2) \tag{1}$$

where x_1 and x_2 are mole fractions, M_1 and M_2 the molar masses, and V_1 and V_2 the molar volumes of 2,3-butanediol and water, respectively. The experimental densities, excess molar volumes, of binary mixtures of 2,3-butanediol (1) + water (2) at different temperatures are listed in Table 2.1. The excess molar volume were correlated by Redlich–Kister (Geyer et al., in press)

$$\Delta E = x_i x_j \sum_{t=0}^{m} A_t (x_i - x_j)^t \tag{2}$$

In this equation, ΔE is the excess property, A_t is a parameter and m is the degree of the polynomial expansion fitting equation (1). The coefficients A_i in the equation (2) were estimated by the least-squares fit method. Values obtained are presented in Table 2.2.

TABLE 2.1 Densities, ρ, excess molar volumes, V^E, for the binary system 1,2 ethandiol (1) + water (2) at different temperatures.

X1	Density	V^E	X1	Density	V^E
T=288.2 K					
0	999.1	0	0.3692	0.6825	−0.478
0.0613	1017.7	−0.211	0.4325	0.7525	−0.385
0.1221	1031.9	−0.435	0.5582	0.8145	−0.296
0.1836	1040.4	−0.593	0.6201	0.8739	−0.206
0.2486	1045	−0.686	0.6825	0.9386	−0.103
0.3083	1046.9	−0.724	0.7525	1	0
T=298.2 K					
0	997	0	0.6822	1036.9	−0.447
0.0605	1013.7	−0.197	0.7446	1035.9	−0.367

TABLE 2.1 *(Continued)*

X1	Density	V^E	X1	Density	V^E
0.1218	1026.3	−0.401	0.8084	1035	−0.283
0.1823	1033.6	−0.541	0.8731	1034	−0.192
0.2437	1037.6	−0.626	0.9354	1033.1	−0.099
0.3051	1039.3	−0.664	1	1032.2	0
T=303.2					
0	994.00	0	0.6825	996.000	1029.8
0.0613	1009.8	−0.197	0.7525	994.000	1028.7
0.1221	1021	−0.384	0.8145	990.100	1027.8
0.1836	1027.6	−0.516	0.8739	986.900	1026.9
0.2486	1031.1	−0.597	0.9386	985.500	1026
0.3083	1008.600	−0.636	1	984.100	1025.2

As can be seen in Figure 2.1 all the binary systems show negative values of the excess molar volume over the whole range of composition, with a maximum at nearly equimolecular composition. From comparison of different 1,2 ethandiol + water binary systems Figure 2.1.

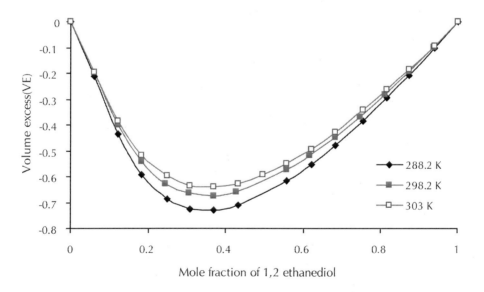

FIGURE 2.1 All the binary systems show negative values of the excess molar volume over the whole range of composition, with a maximum at nearly equimolecular composition.

TABLE 2.2 Regression results for the excess volumes of 2,3-butanediol (1) + water (2) mixtures at various temperatures.

T(K)	A0	A1	A2
308.15	−3.2227	2.4320	0.6711
313.15	−3.200	2.4630	0.5961
318.15	−3.1210	2.5160	0.4902

2.4 CONCLUSION

Densities and excess molar volumes for the binary liquid mixtures of 2,3-butanediol (1) + water (2) were measured at the temperatures of 308.2–318.2 K and atmospheric pressure over the whole range of compositions. Excess molar volumes are negative, are significant. The geometrical models used to predict ternary properties from binary contributions produce in general average absolute deviations lower than 33% for molar excess volume.

KEYWORDS

- **Atmospheric pressure**
- **Binary mixtures**
- **Densimeter**
- **Polynomial equations**
- **Temperatures**

REFERENCES

Dean, J. A. (1985). *Lange's Handbook of Chemistry*, 13th ed., Section 10 and 7, Table no. 10–28 and 7–15, McGraw-Hill, pp. 10–91.

Douheret, G., Pal, A., Hoiland, H., Anowi, O., and Davis, M. I. J. (1991). *Chem. Thermodynamics* **23**, 569–580.

Geyer, H., Ulbig, P., Gornert, M., and Susanto, A. (in press). *J. Chem. Thermodynamics*.

Kapadi, U. R., Hundiwale, D. G., Patil, N. B., Patil, P. R., and Lande, M. K. (2000). *J. Indian Chem. Soc.* **77** 319–321.

Nitta, T., Turek, E. A., Greenkorn, R. A., and Chao, K. C. (1977). *AIChE J.* **23**, 144–160.

Redlich, O. and Kister, A. T. (1948). *Ind. Eng. Chem.* **40**, 345–348.

Span, R. and Wagner, W. (in press). *Int. J. Thermophysics*.

3 A Mathematical Model to Control the Liquid–Liquid Equilibrium Data

CONTENTS

3.1 Introduction 11
3.2 Group Method of Data Handling (GMDH)...12
3.3 The LLE Prediction Using The GMDH-Type Neural Network 14
3.4 Conclusion 25
Keywords 25
References...25

3.1 INTRODUCTION

Liquid-phase equilibrium data of aqueous mixtures with organic solvents play an important role in the design and development of separation processes. In particular, liquid–liquid equilibria (LLE) investigations for ternary mixtures are important in the evaluation of industrial units for solvent extraction processes. The accurate interpretation of phase equilibria of different ternary mixtures is a fundamental and important key to improving solvent extraction techniques (Arce et al., 1995; Senol, 2006; Wu et al., 2003), which can be obtained from direct measurement of LLE data or by the use of different thermodynamic methods (Si-Moussa et al., 2008).

Traditional activity coefficient based thermodynamic models have been successfully used to describe several LLE systems. The nonrandom two-liquid (NRTL) model of Renon and Prausnitz (1968) and the universal quasi-chemical (UNIQUAC) method of Abrams and Prausnitz (1975) models have been used to correlate LLE data for the many multi-component mixtures (Ghanadzadeh et al., 2009; Se and Aznar, 2002), while a group contribution method (UNIFAC) (Fredenslund et. al., 1977) has been widely used to predict the LLE systems.

Recently, new prediction methods were developed using artificial neural networks (ANNs). The ANNs are nonlinear and highly flexible models that have been successfully used in many complicated systems especially in estimating vapor-liquid equilibria (VLE) and LLE data (Reyhani et al., 2009; Sharma et al., 1999; Torrecilla et al., 2008). The ANNs can be considered as universal function approximators. Giving enough data, they can approximate the underlying function with accuracy (Powell, 1987). However, the main disadvantage of traditional neural networks (NNs) is that the detected dependencies are hidden within the NN structure (Nariman-Zadeh and Jamali, 2007).

Conversely, the group method of data handling (GMDH), which was first proposed by Ivakhnenko (1971), is aimed at identifying the functional structure of a model hidden in the empirical data. The main idea of the GMDH is the use of feed-forward networks based on short-term polynomial transfer functions whose coefficients are obtained using regression combined with emulation of the self-organizing activity behind NN structural learning (Farlow, 1984). The GMDH was developed in complex systems for the modeling, prediction, identification, and approximation. It has been shown that, the GMDH is the best optimal simplified model for inaccurate, noisy, or small data sets, with a higher accuracy and a simpler structure than typical full physical models.

In this work, a model for LLE prediction was developed using the GMDH algorithm. The aim of this proposed method is to predict LLE data of a ternary system of [water + acetic acid + 2-ethyl-1-hexanol (2EH)]. Using existing experimental data (36 data set) (Ghanadzadeh et al., 2004) the proposed network was trained. The trained network was used to predict the LLE data in the aqueous and organic phases. Then, the predicted data was compared with the experimental data which have previously reported. In order to investigate the reliability of the proposed method, the accuracy of the model was determined using coefficient of determination (R^2), mean square error (mse), root mean square error (rmse) and mean absolute deviation (MAD).

3.2 GROUP METHOD OF DATA HANDLING (GMDH)

Using the GMDH algorithm, a model can be represented as a set of neurons in which different pairs of them in each layer are connected through a quadratic polynomial and, therefore, produce new neurons in the next layer. Such representation can be used in modeling to map inputs to outputs. The formal definition of the identification problem is to find a function, $\hat{f}$, that can be approximately used instead of the actual one, f, in order to predict output $\hat{y}$ for a given input vector $X = (x_1, x_2, x_3, \cdots, x_n)$ as close as possible to its actual output y. Therefore, given number of observations (M) of multi-input, single output data pairs so that,

$$y_i = f(x_{i1}, x_{i2}, x_{i3}, \cdots, x_{in})(i = 1, 2, 3, \cdots, M) \qquad (1)$$

It is now possible to train a GMDH-type-NN to predict the output values $\hat{y}_i$ for any given input vector $X = (x_{i1}, x_{i2}, x_{i3}, \cdots, x_{in})$, that is

$$\hat{y}_i = \hat{f}(x_{i1}, x_{i2}, x_{i3}, \cdots, x_{in})(i = 1, 2, 3, \cdots, M) \qquad (2)$$

In order to determine a GMDH type-NN, the square of the differences between the actual output and the predicted one is minimized, that is

$$\sum_{i=1}^{M} \left[\hat{f}(x_{i1}, x_{i2}, \ldots, x_i) - y_i \right]^2 \to \min \qquad (3)$$

The general connection between the inputs and the output variables can be expressed by a complicated discrete form of the Volterra functional series (Ivakhnenko, 1971) in the form of

$$y = a_o + \sum_{i=1}^{n} a_i x_i + \sum_{i=1}^{n} \sum_{j=1}^{n} a_{ij} x_i x_j + \sum_{i=1}^{n} \sum_{j=1}^{n} \sum_{k=1}^{n} a_{ijk} x_i x_j x_k + \cdots \qquad (4)$$

where is known as the Kolmogorov–Gabor polynomial (Ivakhnenko, 1971). The general form of mathematical description can be represented by a system of partial quadratic polynomials consisting of only two variables (neurons) in the form of

$$\hat{y} = G(x_i, x_j) = a_o + a_1 x_i + a_2 x_j + a_3 x_i x_j + a_4 x_i^4 + a_5 x_j^2 \cdots \qquad (5)$$

In this way, such partial quadratic description is recursively used in a network of connected neurons to build the general mathematical relation of the inputs and output variables given in equation (4). The coefficients a_i in equation (5) are calculated using regression techniques. It can be seen that a tree of polynomials is constructed using the quadratic form given in equation (5). In this way, the coefficients of each quadratic function G_i are obtained to fit optimally the output in the whole set of input–output data pairs, that is

$$E = \frac{\sum_{i=1}^{M}(y_i - G_i())^2}{M} \to \min \qquad (6)$$

In the basic form of the GMDH algorithm, all the possibilities of two independent variables out of the total n input variables are taken in order to construct the regression polynomial in the form of equation (5) that best fits the dependent observations $(y_i, i = 1, 2, \ldots, M)$ in a least squares sense (Nariman-Zadeh et al., 2002). Using the quadratic sub-expression in the form of equation (5) for each row of M data triples, the following matrix equation can be readily obtained as

$$Aa = Y \qquad (7)$$

where a is the vector of unknown coefficients of the quadratic polynomial in equation (5)

$$a = \{a_o, a_1, a_2, a_3, a_4, a_5\} \qquad (8)$$

$$Y = \{y_1, y_2, y_3 \ldots, y_M\}^T \qquad (9)$$

here Y is the vector of the output's value from observation. It can be readily seen that

$$A = \begin{bmatrix} 1 & x_{1p} & x_{1q} & x_{1p}x_{1q} & x_{1p}^2 & x_{1q}^2 \\ 1 & x_{2p} & x_{2q} & x_{2p}x_{2q} & x_{2p}^2 & x_{2q}^2 \\ \vdots & \vdots & \vdots & \vdots & \vdots & \vdots \\ 1 & x_{Mp} & x_{Mq} & x_{Mp}x_{Mq} & x_{Mp}^2 & x_{Mq}^2 \end{bmatrix} \qquad (10)$$

The least squares technique from multiple regression analysis leads to the solution of the normal equations in the form of

$$a = (A^T A)^{-1} A^T Y \qquad (11)$$

3.3 THE LLE PREDICTION USING THE GMDH-TYPE NEURAL NETWORK

The feed-forward GMDH-type neural network for the ternary system of (water + acetic acid + 2EH) was constructed using an experimental data set which has previously been reported (Ghanadzadeh et al., 2004). The experimental compositions of the mixtures together with the predicted LLE data, using the UNIFAC model, in the aqueous and organic phases at different temperatures are shown in Table 3.1.

TABLE 3.1 Experimental, UNIFAC predicted, and GMDH estimated tie-line data in the aqueous and organic phases for (water (1) + acetic acid (2) + 2EH (3)) at 298.2–313.2 K.

Aqueous phase mole fraction						Organic phase mole fraction					
x_1 (water)			x_2 (acetic acid)			x_1 (water)			x_2 (acetic acid)		
Exp.	UNIFAC	GMDH	Exp.	UNIFAC	GMDH	Exp.	UNIFAC	GMDH	Exp.	UNIFAC	GMDH
T = 298.2 K											
0.9996	0.9997	0.9994	0.0000	0.0000	0.0001	0.1081	0.1181	0.1073	0.0000	0.0000	0.0000
0.9739	0.9735	0.9742	0.0257	0.0264	0.0252	0.1252	0.1265	0.1224	0.0936	0.1495	0.0946
0.9482	0.9444	0.9483	0.0514	0.0555	0.0515	0.1431	0.1453	0.1423	0.1796	0.2092	0.1781
0.8982	0.8934	0.8982	0.1016	0.1065	0.1019	0.1729	0.1777	0.1691	0.2878	0.2822	0.2860
0.8821	0.8782	0.8822	0.1177	0.1216	0.1178	0.1820	0.1874	0.1763	0.3113	0.2990	0.3108
0.8529	0.8527	0.8525	0.1468	0.1471	0.1468	0.1960	0.2035	0.1939	0.3520	0.3230	0.3502
0.8217	0.8239	0.8213	0.1776	0.1737	0.1769	0.2192	0.2271	0.2192	0.3811	0.3511	0.3837
0.7957	0.8000	0.7961	0.2033	0.1951	0.2024	0.2790	0.2834	0.2741	0.4230	0.3944	0.4278
0.6835	0.6870	0.6835	0.2966	0.2899	0.2952	0.3970	0.4035	0.3807	0.4398	0.4225	0.4428
T = 303.2 K											
0.9996	0.9996	0.9996	0.0000	0.0000	0.0001	0.1081	0.1087	0.1071	0.0000	0.0000	0.0000
0.9830	0.9809	0.9827	0.0168	0.0190	0.0173	0.1269	0.1254	0.1354	0.0934	0.1211	0.0917
0.9483	0.9489	0.9482	0.0516	0.0507	0.0518	0.1459	0.1450	0.1487	0.1868	0.2015	0.1854
0.8986	0.8986	0.8990	0.1013	0.1014	0.1013	0.1763	0.1764	0.1717	0.2868	0.2902	0.2878
0.8826	0.8811	0.8831	0.1172	0.1188	0.1170	0.1848	0.1854	0.1783	0.3107	0.3096	0.3125
0.8533	0.8505	0.8538	0.1463	0.1493	0.1458	0.1995	0.2011	0.1935	0.3476	0.3387	0.3509
0.8224	0.8221	0.8211	0.1769	0.1776	0.1775	0.2116	0.2168	0.2230	0.3969	0.3629	0.3898
0.7962	0.7976	0.7955	0.2027	0.1999	0.2031	0.2640	0.2677	0.2705	0.4350	0.4164	0.4304
0.6960	0.6577	0.6962	0.2916	0.3135	0.2927	0.3731	0.3787	0.3738	0.4788	0.4440	0.4681
T = 308.2 K											
0.9996	0.9996	0.9997	0.0000	0.0000	0.0000	0.1081	0.1087	0.1070	0.0000	0.0000	0.0003
0.9811	0.9849	0.9813	0.0186	0.0112	0.0188	0.1285	0.1266	0.1323	0.0933	0.1262	0.0933
0.9484	0.9546	0.9487	0.0515	0.0394	0.0514	0.1492	0.1479	0.1501	0.1864	0.2068	0.1878

A Mathematical Model to Control the Liquid–Liquid Equilibrium Data

TABLE 3.1 *(Continued)*

\multicolumn{5}{l}{Aqueous phase mole fraction}					\multicolumn{5}{l}{Organic phase mole fraction}				

\multicolumn{3}{l}{x_1 (water)}			\multicolumn{3}{l}{x_2 (acetic acid)}			\multicolumn{3}{l}{x_1 (water)}			\multicolumn{3}{l}{x_2 (acetic acid)}		

Exp.	UNIFAC	GMDH	Exp.	UNIFAC	GMDH	Exp.	UNIFAC	GMDH	Exp.	UNIFAC	GMDH
0.8989	0.9071	0.8997	0.1009	0.0845	0.1006	0.1797	0.1797	0.1733	0.2863	0.2898	0.2908
0.8830	0.8917	0.8840	0.1167	0.0992	0.1162	0.1881	0.1887	0.1798	0.3099	0.3079	0.3156
0.8538	0.8634	0.8548	0.1458	0.1263	0.1449	0.2029	0.2048	0.1953	0.3468	0.3355	0.3553
0.8232	0.8335	0.8231	0.1761	0.1547	0.1762	0.2360	0.2206	0.2372	0.4027	0.3582	0.4042
0.7968	0.8058	0.7976	0.2001	0.1810	0.2011	0.2730	0.2784	0.2664	0.4398	0.4131	0.4326
0.6750	0.6837	0.6755	0.3128	0.2923	0.3125	0.3967	0.4023	0.4005	0.4679	0.4488	0.4773
\multicolumn{12}{l}{T = 313.2 K}											
0.9996	0.9993	0.9998	0.0000	0.0000	0.0000	0.1081	0.1140	0.10707	0.0000	0.0000	0.0008
0.9793	0.9526	0.9798	0.0205	0.0473	0.0200	0.1301	0.1311	0.129953	0.0929	0.1488	0.0965
0.9486	0.9214	0.9478	0.0513	0.0785	0.0518	0.1280	0.1535	0.134169	0.1858	0.2267	0.1782
0.8992	0.8946	0.8997	0.1006	0.1053	0.1002	0.1420	0.1671	0.140987	0.2700	0.2634	0.2702
0.8835	0.8740	0.8823	0.1162	0.1259	0.1175	0.1685	0.1775	0.180179	0.3320	0.2874	0.3221
0.8508	0.8513	0.8504	0.1487	0.1482	0.1489	0.1820	0.1997	0.190435	0.3600	0.3293	0.3598
0.8319	0.8333	0.8313	0.1675	0.1652	0.1674	0.2022	0.2117	0.211123	0.3830	0.3477	0.3867
0.7661	0.7652	0.7670	0.2243	0.2238	0.2246	0.2579	0.2614	0.260131	0.4150	0.4017	0.4166
0.6815	0.7017	0.6808	0.2896	0.2685	0.2901	0.3770	0.3790	0.387344	0.4190	0.4185	0.4165

This data set consists of 36 points for four different temperatures. The data was divided into two parts: 70% (26 points) was used as training data, and 30% (10 points) was used as test data. The three feed fractions in the compositions and temperature were used as inputs of the GMDH-type network. The three mole fractions in the aqueous phase (x_{11}, x_{21}, x_{31}) and three mole fractions in the organic phase (x_{13}, x_{23}, x_{33}) were used as desired outputs of the neural network.

In order to estimate the mole fractions in the aqueous and organic phases, using the GMDH-type network, six polynomial equations were obtained (Table 3.2). In this table, z_1 is the temperature and z_2, z_3, and z_4 are the normalized feed fractions of water acetic acid and the organic solvent (2EH), respectively, in the overall composition. The proposed model was used to calculate the mole fractions (the output data) of the components in the aqueous and organic phases. The calculated values are also presented in Table 3.1.

TABLE 3.2 Polynomial equations of the GMDH model for the system (water (1) + acetic acid (2) + 2EH (3)).

x_{11} (mole fraction of water in aqueous phase)

$Y_1 = 0.6665 + 2.8629 z_2 - 2.1985 z_4 - 1.2511 z_2^2 - 8.2906 z_4^2 + 4.4163 z_2 z_4$

$Y_2 = 0.00004 + 0.0060 z_1 + 0.00005 z_4 - 0.00001 z_1^2 + 0.000006 z_4^2 + 0.0147 z_1 z_4$

$Y_3 = 0.8780 - 20.4823 z_2 + 17.8765 y_1 + 115.1454 z_2^2 + 77.4637 y_1^2 - 188.9360 z_2 y_1$

TABLE 3.2 *(Continued)*

$Y_4 = 0.1182 + 1.1226z_3 + 1.8754y_2 - 1.4610z_3^2 - 0.9900y_2^2 - 2.0631z_3y_2$

$x_{11} = 0.0041 + 0.5801y_3 + 0.4115y_4 - 178.5454y_3^2 - 178.6082y_4^2 + 357.1579y_3y_4$

x_{21} (mole fraction of acid in aqueous phase)

$Y_1 = -0.0005 + 0.0042z_1 - 0.0280z_2 - 0.000005z_1^2 - 0.3670z_2^2 - 0.0018z_1z_2$

$Y_2 = -0.0400 + 1.0015z_3 + 0.3367z_4 + 0.1694z_3^2 + 1.2172z_4^2 - 0.4800z_3z_4$

$Y_3 = 0.0367 + 0.8689y_1 - 0.2970z_4 + 0.3325y_1^2 - 1.2173z_4^2 - 1.5382y_1z_4$

$Y_4 = 0.0002 - 0.1030z_3 + 1.0972y_2 + 11.5102z_3^2 + 12.7323y_2^2 - 24.1778z_3y_2$

$x_{21} = 0.00009 + 0.0393y_3 + 0.9602y_4 - 40.2470y_3^2 - 42.2506y_4^2 + 82.4972y_3y_4$

x_{31} (mole fraction of 2EH in aqueous phase)

$Y_1 = 10.6417 - 23.2880z_2 - 16.5302z_3 + 12.7808z_2^2 + 6.4240z_3^2 + 18.0832z_2z_3$

$Y_2 = -0.0000007 - 0.0001z_1 - 0.000001z_4 + 0.0000004z_1^2 - 0.0000001z_4^2 - 0.0004z_1z_4$

$Y_3 = -0.0007 + 0.9644y_1 + 0.0185z_4 - 10.0050y_1^2 - 0.1412z_4^2 + 7.0469y_1z_4$

$Y_4 = 0.2469 - 50.2498y_2 - 6.6085z_4 + 2816.0600y_2^2 + 43.8112z_4^2 + 660.5264z_4y_2$

$x_{31} = -0.00005 + 1.3737y_3 - 0.0909y_4 - 4.0010y_3^2 + 12.8576y_4^2 - 41.7301y_3y_4$

x_{13} (mole fraction of water in organic phase)

$Y_1 = 1.4920 - 6.7541z_4 - 26.7141z_3 + 11.2604z_4^2 + 125.6225z_3^2 + 62.2677z_3z_4$

$Y_2 = 10.0400 - 224.5310z_2 - 169.0174z_3 + 125.6225z_2^2 + 74.6152z_3^2 + 188.9773z_2z_3$

$Y_3 = 0.00001 + 0.0019z_1 - 0.00003z_4 - 0.000002z_1^2 - 0.000004z_4^2 - 0.0114z_1z_4$

$Y_4 = -0.3236 + 2.4857y_1 + 5.0852z_4 - 1.8345y_1^2 - 22.6687z_4^2 - 9.3191y_1z_4$

$Y_5 = -0.0397 + 1.4501y_2 + 0.0046y_3 - 1.8499y_2^2 - 1.9058y_3^2 + 2.6883y_2y_3$

$x_{13} = 0.0004 + 17.9130y_4 - 16.9172y_5 - 38.5793y_4^2 + 38.5890y_5^2 - 0.0009y_4y_5$

x_{23} (mole fraction of acid in organic phase)

$Y_1 = 0.00001 + 0.0018z_1 - 0.00008z_4 + 0.000001z_1^2 - 0.000009z_4^2 - 0.0249z_1z_4$

$Y_2 = 0.3605 + 0.9860z_3 - 3.0302z_4 - 1.5247z_3^2 - 10.9551z_4^2 + 4.3184z_3z_4$

$Y_3 = 0.00001 + 0.0018z_1 - 0.00008z_4 + 0.000001z_1^2 - 0.000009z_4^2 - 0.0249z_1z_4$

$Y_4 = 0.3085 - 0.0275y_1 + 1.2221z_2 - 0.2650y_1^2 - 1.3901z_2^2 + 0.9673y_1z_2$

$Y_5 = -0.0056 + 0.6816y_2 + 0.3783y_3 + 3.4771y_2^2 + 3.0340y_3^2 - 6.6354y_2y_3$

$x_{23} = 0.0008 - 0.8891y_4 + 1.8710y_5 + 63.8928y_4^2 + 59.4812y_5^2 - 123.3386y_4y_5$

x_{33} (mole fraction of 2EH in organic phase)

$Y_1 = 0.0076 - 0.0069z_1 + 0.4237z_2 + 0.00004z_1^2 + 5.5472z_2^2 - 0.0209z_1z_2$

$Y_2 = -85.7756 + 199.5904z_3 + 138.7723z_4 - 114.6674z_3^2 - 57.8168z_4^2 - 162.7486z_3z_4$

$Y_3 = -0.0080 + 0.2309y_1 + 0.8019y_2 + 1.1985y_1^2 + 1.5978y_2^2 - 2.8250y_2y_1$

$Y_4 = -0.8526 + 29.7443z_4 + 5.7680z_3 - 114.6674z_4^2 - 9.7356z_3^2 - 66.5861z_3z_4$

$x_{33} = -0.0029 + 1.8385y_3 - 0.8294y_4 + 67.8869y_3^2 + 70.2118y_4^2 - 138.1058y_3y_4$

The developed GMDH neural network was successfully used to obtain six models for calculation of the LLE. The optimal structures of the developed neural network with 2-hidden layers are shown in Figures 3.1 and 3.2. For instance, "bbbdccad" and

A Mathematical Model to Control the Liquid–Liquid Equilibrium Data 17

"dcddbcad" are corresponding genome representations for the mole fractions of water in the aqueous and organic phases, respectively. In which, a, b, c, and d stand for temperature and feed fractions (water, acetic acid, and 2EH), respectively. All input variables were accepted by the models. In other words, the GMDH-type-NN provides an automated selection of essential input variables, and builds polynomial equations for the LLE modeling. These polynomial equations show the quantitative relationship between input and output variables (Table 3.2).

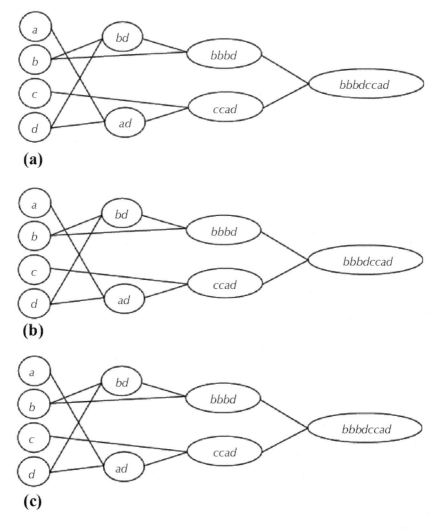

FIGURE 3.1 Developed structure of GMDH-type-NN model for the ternary system in aqueous phase; A) x_{11}, B) x_{21}, C) x_{31}.

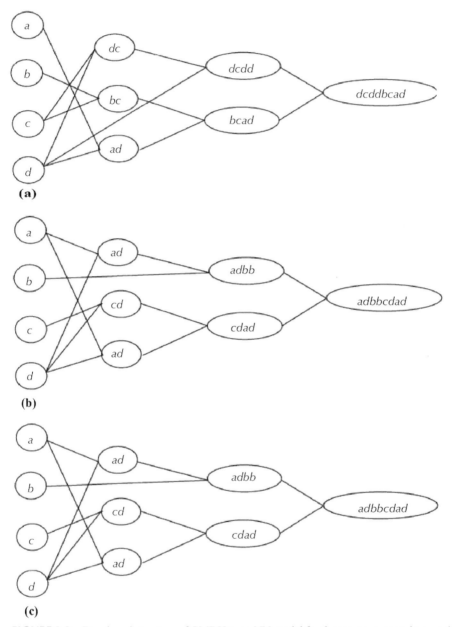

FIGURE 3.2 Developed structure of GMDH-type-NN model for the ternary system in organic phase, A) x_{13}, B) x_{23}, C) x_{31}.

Our proposed models behavior in prediction of the LLE is demonstrated in Figures 3.3 and 3.4. The results of the developed models give a close agreement between observed and predicted values of the LLE. Some statistical measures are given in Table

3.3, in order to determine the accuracy of the models. These statistical values are based on R^2 as absolute fraction of variance, rmse as root-mean squared error, mse as mean squared error, and MAD as mean absolute deviation which are defined as follows:

$$R^2 = 1 - \left[\frac{\sum_{i=0}^{M}\left(Y_{i(model)} - Y_{i(acutual)}\right)^2}{\sum_{i=1}^{M}\left(Y_{i(acutual)}\right)^2} \right] \quad (12)$$

$$RMSE = \left[\frac{\sum_{i=0}^{M}\left(Y_{i(model)} - Y_{i(acutual)}\right)^2}{M} \right]^{1/2} \quad (13)$$

$$MSE = \frac{\sum_{i=0}^{M}\left(Y_{i(model)} - Y_{i(acutual)}\right)^2}{M} \quad (14)$$

$$MAD = \frac{\sum_{i=0}^{M}\left|Y_{i(model)} - Y_{i(acutual)}\right|}{M} \quad (15)$$

TABLE 3.3 Model statistics and information for the group method of data handling-type neural network model for predicting LLE data.

Mole fraction	Statistic	R^2	rmse	mse	MAD
Aqueous rich-phase					
x_{11}	Training	1.0000	0.00052	0.00000	0.00041
	Testing	0.9999	0.00065	0.00000	0.00055
x_{21}	Training	0.99999	0.00049	0.00000	0.00000
	Testing	0.99998	0.00069	0.00000	0.00000
x_{31}	Training	0.99600	0.00045	0.00000	0.00031
	Testing	0.98030	0.00078	0.00000	0.00061
Organic rich-phase					
x_{13}	Training	0.99904	0.00629	0.00003	0.00481
	Testing	0.99939	0.00599	0.00003	0.00501
x_{23}	Training	0.99987	0.00336	0.00001	0.00003
	Testing	0.99966	0.00652	0.00004	0.00004
x_{33}	Training	0.99995	0.00407	0.00002	0.00266
	Testing	0.99979	0.00709	0.00005	0.00462

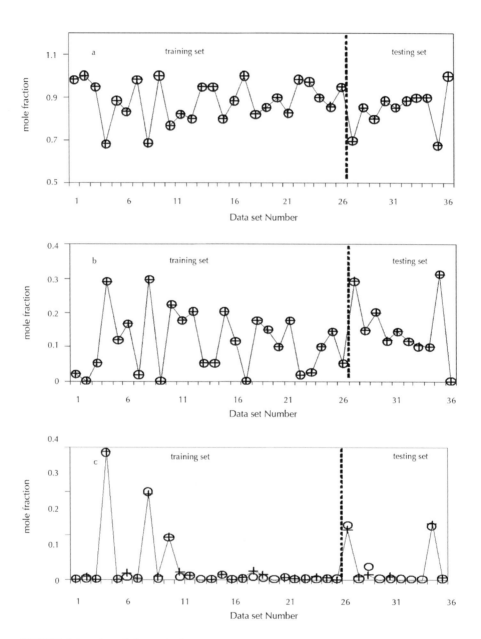

FIGURE 3.3 Plot of the mole fractions against data set number in the aqueous phase to illustrate the prediction of the experimental data for (Water + Acetic Acid + 2EH) using the GMDH model; (○) experimental points; (+) calculated points.

A Mathematical Model to Control the Liquid–Liquid Equilibrium Data 21

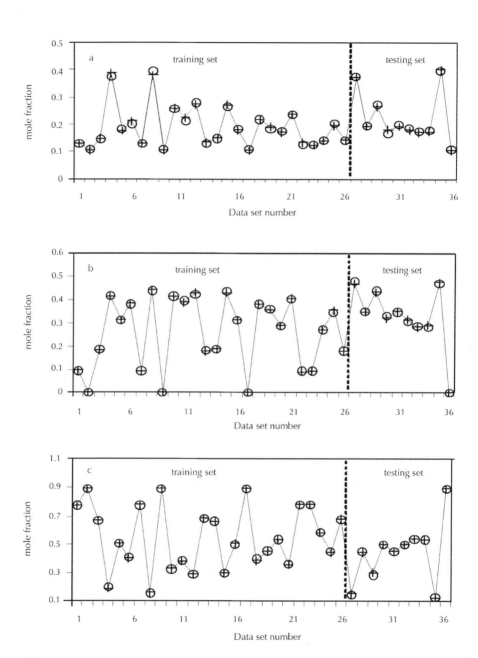

FIGURE 3.4 Plot of the mole fractions against data set number in the organic phase to illustrate the prediction of the experimental data for (Water + Acetic Acid + 2EH) using the GMDH model; (○) experimental points; (+) calculated points.

After the learning and training with the GMDH, the output (the mole fractions) is generated. The output data of the models are given in Table 3.1. Now, we can compare the experimental data and those predicted by the UNIFAC model, which has been reported in our earlier publication (Ghanadzadeh et al., 2004), with the output of GMDH (see Table 3.1).

The root-mean-square deviation (rmsd) can be taken to be a measure of the precision of the correlations. The rmsd value was calculated from the difference between the experimental and calculated mole fractions according to the following equation,

$$RMSD = \sqrt{\frac{\sum_{k=1}^{n}\sum_{j=1}^{2}\sum_{i=1}^{3}\left(x_{ijk}-\hat{x}_{ijk}\right)^2}{6n}} \tag{16}$$

where n is the number of tie-lines, x indicates the experimental mole fraction, $\hat{x}$ the calculated mole fraction, and subscript I indexes components, j indexes phases and the subscript i indexes components, j indexes phase and k = 1, 2, ..., n (tie-lines).

The rmsd is a measure of the agreement between the experimental and calculated data. The rmsd for the ternary system at each temperature was listed in Table 3.4. A comparison between the rmsd values using the proposed GMDH and the UNIFAC models were demonstrated in this table. As it can be seen, the former model gives a better agreement with the experimental data respect to those previously obtained using the UNIFAC.

TABLE 3.4 Mean deviations [rmsd%] in different models at various temperatures.

T/K	UNIFAC	GMDH
298.2	1.75	0.31
303.2	1.82	0.35
308.2	2.06	0.42
313.2	2.78	0.28

To indicate the ability of the solvent in the recovery of acetic acid, distribution coefficients for water ($D_1 = x_{13}/x_{11}$) and acetic acid ($D_2 = x_{23}/x_{21}$) and separation factors ($S = D_2/D_1$) were calculated from experimental and predicted data. The x_{13} and x_{23} are the mole fractions of water and acetic acid in organic-rich phase, respectively. The x_{11} and x_{21} are the mole fractions of water and the acid in aqueous phase, respectively. The distribution coefficients and separation factors for the ternary system are listed in Table 3.5.

TABLE 3.5 Experimental and calculated separation factors (S) and distribution coefficients of acetic acid (D2) at different temperatures.

Experimental		UNIFAC		GMDH	
D2	S	D2	S	D2	S
T = 298.2 K					
3.64	28.33	5.66	43.58	3.75	29.88

TABLE 3.5 *(Continued)*

Experimental		UNIFAC		GMDH	
D2	S	D2	S	D2	S
3.49	23.15	3.77	24.50	3.46	23.05
2.83	14.72	2.65	13.32	2.81	14.91
2.64	12.82	2.46	11.52	2.64	13.20
2.40	10.43	2.20	9.20	2.39	10.49
2.15	8.04	2.02	7.33	2.17	8.13
2.08	5.93	2.02	5.71	2.11	6.14
1.48	2.55	1.46	2.48	1.50	2.69
T = 303.2 K					
5.56	43.07	6.37	49.86	5.30	38.47
3.62	23.53	3.97	26.01	3.58	22.82
2.83	14.43	2.86	14.58	2.84	14.88
2.65	12.66	2.61	12.39	2.67	13.23
2.38	10.16	2.27	9.59	2.41	10.62
2.24	8.72	2.04	7.75	2.20	8.09
2.15	6.47	2.08	6.21	2.12	6.23
1.64	3.06	1.42	2.46	1.60	2.98
T = 308.2 K					
5.02	38.30	11.27	87.66	4.96	36.81
3.62	23.01	5.25	33.88	3.65	23.09
2.84	14.19	3.43	17.31	2.89	15.01
2.66	12.47	3.10	14.67	2.72	13.35
2.38	10.01	2.66	11.20	2.45	10.73
2.29	7.98	2.32	8.75	2.29	7.96
2.20	6.41	2.28	6.61	2.15	6.44
1.50	2.55	1.54	2.61	1.53	2.58
T = 313.2 K					
4.53	34.11	3.15	22.86	4.83	36.38
3.62	26.84	2.89	17.33	3.44	24.30
2.68	17.00	2.50	13.39	2.70	17.21
2.86	14.98	2.28	11.24	2.74	13.42
2.42	11.32	2.22	9.47	2.42	10.79
2.29	9.41	2.10	8.28	2.31	9.10
1.85	5.50	1.79	5.25	1.85	5.47
1.45	2.62	1.56	2.89	1.44	2.52

For the investigated system, the experimental and calculated data, using the UNIFAC and GMDH models, indicate that 2EH has a high separation factor (separation factor varying between 3.06 and 43.07 at 303.2 K). Figure 3.5 presents the experimental and estimated distribution coefficients of the acid. The variation of separation factor of the acid as a function of the mole fraction of the solute in the organic phase for the ternary system was shown in Figure 3.6. As it can be seen, the GMDH model gives excellent agreement with the experimental data respect to those obtained using the other model.

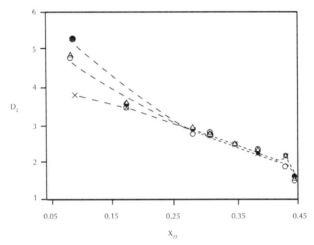

FIGURE 3.5 Plot of the distribution coefficient (D_2) of acetic acid as a function of the mole fraction (X_{23}) of acetic acid in the organic phase at different temperatures; (×) 298.2 K, (●) 303.2 K, (Δ) 308.2 K, (○) 313.2 K.

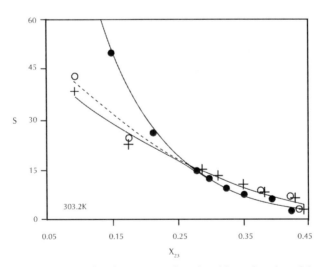

FIGURE 3.6 Plot of the separation factors (S) of acetic acid as a function of the mole fraction (X_{23}) of acetic acid in the organic phase at 303.2 K; (○) experimental points; (+) GMDH calculated points (●) UNIFAC calculated.

3.4 CONCLUSION

In this study, a feed-forward GMDH-type neural network model was developed using experimental LLE data for the (water + acetic acid + 2-ethyl-1-hexanol) ternary system over the temperature range of 298.2–313.2 K. The LLE data were predicted by the GMDH model and the results compared with the experimental data and also those previously obtained using the UNIFAC model. Despite the complexity of the system studied, the GMDH model permits a good prediction of the phase equilibrium. Thus, the GMDH model is suitable for predicting the LLE data. The agreements between the experimental and calculated data were found to be excellent.

For the investigated system, the average RMSD values between the observed and calculated mole fractions using the UNIFAC and GMDH models were 2.10 and 0.34%, respectively. As it can be seen, the level of deviations indicates the superiority of GMDH-type neural network as the preferred predicting model for the system considered.

The separation factor and distribution coefficient for the organic solvent used in this work were calculated by both the UNIFAC and GMDH models. The data indicate that 2-ethyl-1-hexanol (2EH) has relatively high separation factor, which show the ability of this solvent to extract acetic acid from water. It might be concluded that 2EH may serve as an adequate solvent for the extraction of the acid from aqueous solution.

KEYWORDS

- **Artificial neural networks**
- **Group method of data handling**
- **Liquid-liquid equilibria**
- **Mole fraction**
- **Neural networks**
- **Root mean square deviation**

REFERENCES

Abrams, D. S., and Prausnitz, J. M. (1975). Statistical thermodynamics of liquid mixtures: a new expression for excess Gibbs energy of partly or completely miscible systems. *AIChE J.* **21**, 116–128.

Arce, A., Blanco, A., Martinez-Ageitos, J., and Vidal, I. (1995). LLE data for the systems water + (methanol or ethanol) + *n*-amyl acetate. *Fluid Phase Equilib.* **109**, 291–297.

Farlow, S. J. (1984). *Self-organizing method in modelling: GMDH-type algorithm.* Marcel Dekker. New York.

Fredenslund, A., Jones, R. L., and Prausnitz, J. M., (1975). Group contribution estimation of activity coefficients in non-ideal solutions. *AIChE J.* **21**, 1086–1099.

Ghanadzadeh, H., Ghanadzadeh, A., and Sariri, R. (2004). Liquid-liquid equilibria for water + acetic acid + 2-ethyl-1-hexanol: Experimental data and prediction. *J. Chem. Thermodyn.* **36**, 1001–1006.

Ghanadzadeh, A., Ghanadzadeh, H., and Bahrpaima, Kh. (2009). Experimental and theoretical study of the phase equilibria in ternary aqueous mixtures of 1,4-butanediol with alcohols at 298.2 K. *J. Chem. Eng. Data* **54**, 1009–1014.

Ivakhnenko, A. G. (1971). Polynomial theory of complex systems. *IEEE Trans. Syst. Man Cybern.* **1**, 364–78.

Nariman-Zadeh, N., Darvizeh, A., Felezi, M. E., and Gharababaei, H. (2002). Polynomial modeling of explosive compaction process of metallic powders using GMDH-type neural networks and singular value decomposition. *Mater. Sci. Eng.* **10**, 727–744.

Nariman-Zadeh, N., and Jamali, A. (2007). Pareto genetic design of GMDH-type neural networks for nonlinear systems. *In proceedings of the international workshop on inductive modeling.* J. Drchal and J. Koutnik (Eds.). Czech Technical University, Prague, Czech Republic, pp.96–103.

Powell, M. J. D. (1987). *Radial basis functions for multivariable interpolation: A review in algorithms for approximation.* J. C. Mason and M. G. Cox (Eds.). Clarendon Press, Oxford, UK, pp. 143–167.

Renon, H. and Prausnitz, J. M. (1968). Local compositions in thermodynamic excess functions for liquid mixtures. *AIChE J.* **14**, 135–144.

Reyhani, S. Z, Ghanadzadeh, H., Puigjaner, L., and Recances, F. (2009). Estimation of liquid–liquid equilibrium for a quaternary system using the GMDH algorithm. *Ind. Eng. Chem. Res.* **48**, 2129–2134.

Se, R. A. G. and Aznar, M. (2002). Thermodynamic modelling of phase equilibrium for water + poly(ethylene glycol) + salt aqueous two-phase systems. *Brazilian J. Chem. Eng.* **19**, 255–266.

Senol, A. (2006). Liquid–liquid equilibria for the system (water + carboxylic acid + chloroform): Thermodynamic modeling. *Fluid Phase Equilib.* **243**, 51–56.

Sharma, R., Singhal, D., Ghosh, R., and Dwivedi, A. (1999). Potential applications of artificial neural networks to thermodynamics: Vapor-liquid equilibrium predictions. *Comput. Chem. Eng.* **23**, 385–390.

Si-Moussa, C., Hanini, S., Derriche, R., Bouhedda, M., and Bouzidi, A. (2008). Prediction of high-pressure vapor liquid equilibrium of six binary systems, carbon dioxide with six esters, using an artificial neural network model. *Brazilian J. Chem. Eng.* **25**, 183–199.

Torrecilla, J. S., Deetlefs M., Seddon, K. R., and Rodríguez, F. (2008). Estimation of ternary liquid–liquid equilibria for arene/alkane/ionic liquid mixtures using neural networks. *Phys. Chem. Chem. Phys.* **10**(33), 5114–5120.

Wu, C. T., Marsh, K. N., Deev, A. V., and Boxall, J. A. (2003). Liquid–liquid equilibria of room-temperature ionic liquids and butan-1-ol. *J. Chem. Eng. Data* **48**, 486–491.

4 Computer Calculations for Multicomponent Vapor–Liquid and Liquid–Liquid Equilibria

CONTENTS

4.1	Introduction	27
4.2	Experimental	29
	4.2.1 Materials	29
	4.2.2 Apparatus and Procedure	29
4.3	Discussion and Results	30
4.4	Conclusion	40
Keywords		40
References		40

NOMENCLATURES

a = Optimized interaction parameter
A, B, and C = Antoine equation parameters
B_{ij} = Second virial coefficient
C = Number components
NRTL = Non random two liquid
q_i = Relative surface area per molecule
r_i = Number of segments per molecule
rmsd = Root mean square deviation %
T = Absolute temperature (Kelvin)
u_{ij} = Interaction energy
Uniquac = Universal quasi chemical
x = Mole fraction
x_i = Equilibrium mole fraction of component i

4.1 INTRODUCTION

The precise vapor–liquid equilibrium (VLE) data of binary mixtures like alcohol–alcohol are important to design many chemical processes and separation operations. The

VLE investigations of binary systems have been the subject of much interest in recent years (Artigas et al., 1997, 2001; Rodriguez et al., 2002; Seo et al., 2000). In the past, several authors have reported isobaric VLE data for the system of *tert*-butanol (TBA) + water (Hirata et al., 1975; Kojima and Tochigi, 1979; Quitzsch et al., 1968; Suska et al., 1970). Darwish and Al-Anber (1997) have presented isobaric VLE data for the binary systems of TBA + water and of TBA + isobutanol at 94.9 kPa. We have recently reported (Ghanadzadeh and Ghanadzadeh, 2002a) the LLE data for ternary mixture of water + TBA + 2-ethyl-1-hexanol, where only one liquid pair of partially miscible (2-ethyl-1-hexanol + water) and two liquid pairs of completely miscible (water + TBA) and (2-ethyl-1-hexanol + TBA). From the experimental results, 2-ethyl-1-hexanol was chosen as the best solvent for recovering TBA and n-butanol (NBA) from aqueous solution (Ghannadzadeh, 1993a; Ghanadzadeh and Ghanadzadeh, 2002a, 2003a).

In order to know the activity coefficients of the binary systems of TBA + 2-ethyl-1-hexanol and NBA + 2-ethyl-1-hexanol, the VLE data are also needed. These experimental data are determined using a method based on gas chromatographic technique described previously (Ghannadzadeh, 1993a). The experimental activity coefficients and VLE data can be correlated to several equilibrium methods such as universal quasi-chemical (UNIQUAC) and NTRL models. Equilibrium models, such as the UNIQUAC model (Renon and Prausnitz, 1968) and the non-random two-liquid model (NRTL) (Abrams and Prausnitz, 1975) have been successfully applied for the correlation of several liquid–liquid and vapor–liquid systems. These models depend on optimized interaction parameters between each pair of components in the systems, which can be obtained by experiments.

In this research we have presented experimental activity coefficients and isobaric VLE data for the binary systems TBA + 2-ethyl-1-hexanol and NBA + 2-ethyl-1-hexanol at 101.3 kPa, that have not been reported in literature. The calculated activity coefficients have been correlated by fitting the binary interaction parameters of different models UNIQUAC (Renon and Prausnitz, 1968) and NRTL (Abrams and Prausnitz, 1975). The precise liquid–liquid equilibria (LLE) data is necessary to rational design of many chemical processes and optimize extraction processes. Many researchers have investigated various kinds of multi-component systems in order to understand and provide further information about the phase behavior and the thermodynamic properties of such systems (Artigas et al., 2001; Hirata et al., 1975; Kojima and Tochigi, 1979; Quitzsch et al., 1968; Rodriguez et al., 2002; Seo et al., 2000; Suska et al., 1970). In order to be able to predict LLE in multi-component systems an adequate equilibrium model should be presented earlier researchers reported the correlation of LLE systems with the solution model of the UNIQUAC (Darwish and Al-Anber, 1997; Ghanadzadeh and Ghanadzadeh, 2002a). This model could depends on optimized interaction parameters between each pair of components in the system, which can be carried out by experiments. By optimizing the interaction parameter, the UNIQUAC equation can be best fitted to the experimental composition.

In recent years, there has been increasing attraction in adding a range of oxygenated compounds, mainly alcohols and ethers, to gasoline due to their octane enhancing (Ghanadzadeh and Ghanadzadeh, 2003a). Also, by using oxygenated compounds instead of lead in the gasoline the levels of contamination can be remarkably reduced. In some countries, the oxygenated compounds such as, methyl *tert*-butyl ether (MTBE), *tert*-amyl methyl ether (TAME) and *tert*-amyl alcohol (TAOH) have been used. Methanol is one of the most appropriated oxygenated compounds for this purpose because of its physical–chemical properties. Methanol can be easily produced from a variety of organic materials (Ghannadzadeh, 1993a), petroleum, and coal. However, phase separation and the high vapor pressure of methanol in gasoline had been a restriction for achieving a wide application. Therefore, thermodynamic studies and the precise LLE data for methanol and representative compounds of the gasoline is necessary in order to the determine region of solubility of methanol and plait point of the interest system.

Recently, Trejo et al. (Renon and Prausnitz, 1968) have reported LLE measurements for methanol and representative compounds of the gasoline, and their investigation were somehow important in gasoline reformation with methanol. However, the present study is an effort to show experimentally that methanol can be used as an appropriate oxygenated compound in gasoline formulations. In view of this, we will apply for the first time, the liquid–liquid phase equilibria data for three different ternary systems, cyclohexane + methanol + ethyl benzene at 288.2 K. Where the paraffin is cyclohexane a representative component of the gasoline, methanol, is the oxygenated compound, and the aromatic hydrocarbons are benzene, ethyl benzene. A high aromatic gasoline (35.4 vol% aromatic, 60.4 vol% saturates, and 4.2 vol% olefins) having density of 0.738 g/ml was used in this study. The UNIQUAC model was used to correlate the experimental LLE data. However, the values for the interaction parameters were observed for the UNIQUAC model. The influence of aromatic compounds on mutual solubility of cyclohexane and methanol was also investigated at 288.2 K.

4.2 EXPERIMENTAL

4.2.1 Materials

The TBA, NBA and 2-ethyl-1-hexanol were obtained from Merck with purity better than 99% and were used without further purification. The purity of these materials was checked by gas chromatography (GC).

4.2.2 Apparatus and Procedure

The VLE measurements were carried out using a Labodest unit built by Fischer, and equipped with a Cottrell pump. The apparatus and procedure have been described completely by Artigas (Artigas et al., 1997) and Walas (1985). The temperature of the cell was maintained with an accuracy of within ± 0.1 K. The composition analysis of the vapor and liquid phase were determined using Konik GC equipped with a thermal conductivity detector (TCD) and Shimadzu C-R2AX integrator. The column, injector and detector temperatures were 493.2, 404.2, and 433.2 K, respectively. A 2 m 2 mm column was used to separate the components. The liquid–liquid phase equilibria measurements under ambient pressure and temperature (288.2 K) were carried out using an apparatus of 300 ml glass cell. The temperature of the cell was controlled by a water jacket and measured with a copper-constantan thermocouple and was estimated to be accurate within ± 0.1 K. A series of LLE measurements were performed by changing the composition of the mixture.

The prepared mixtures were placed in the extraction vessel, and stirred for 2 hr and then left to settle for 4 hr. Samples were taken by a syringe (Gaschromatographic's Hamilton 0.4 µl) from both the upper (cyclohexane) phase and lower layers (aromatic phase). Both phases were analyzed using Konik GC equipped with a TCD and Shimadzu C-R2AX integrator. A 2 m × 2 mm column was used to separate the components

4.3 DISCUSSION AND RESULTS

The experimental and predicted activity coefficients of the components, γ_i, are compiled in Tables 4.1–4.2. The same data for the binary systems are graphically represented in Figures 4.1 and 4.2. The experimental activity coefficients of the components, γ_i, were obtained using the GC method (Ghannadzadeh, 1993a). The activity coefficients can be expressed in function of the areas measured in the chromatogram, in the following form:

$$\gamma_i = \frac{A_i}{x_i A_i'} \left\{ \frac{\exp\left[\left(\frac{2}{V}\right)\sum_j^m n_j B_{ij}\right]}{\exp\left[\left(\frac{2}{V}\right)\sum_j^m n_j B_{ij}\right]} \right\} \quad (1)$$

where V is volume of the sample, n_i and n_j are the numbers of moles of the component i and j, respectively, B_{ij} is the second virial coefficient, and A' is standard area, A_i is the area of the pick of component i.

As the exponential terms of the equation 1 spread to the unit, equation (1) reduces to

$$\gamma_i = \frac{A_i}{x_i A_i'} \quad (2)$$

The UNIQUAC and NTRL models were used to correlate the experimental activity coefficients and VLE data. As it can be seen from Figures 4.1 and 4.2, the experimental data (solid line) are in good agreement with the predicted data (dashed line). Although the both equations adequately fit the experimental data for these binary systems, however, the UNIQUAC model fit better the experimental data. The non-ideal behavior of the both systems is quiet obvious in these diagrams. A positive deviation is observed for the systems as indicated by the positive values of activity coefficients over the composition range of 0.1–0.9.

TABLE 4.1 Experimental and predicted activity coefficient for the binary system of TBA + 2-ethyl-1-hexanol.

Activity coefficient TBA			Activity coefficient 2-ethyl-1-hexanol		
Experimental	UNIQUAC	NRTL	Experimental	UNIQUAC	NRTL
4.7600	4.3899	4.2981	1.0000	1.0000	1.0000
2.2834	3.3237	2.3596	1.1700	1.0250	1.1070
2.1000	2.5914	2.0012	1.1900	1.0500	1.1120
2.3990	1.7175	1.9945	1.3000	1.2538	1.2290
1.2698	1.2759	1.1980	2.0600	1.8976	1.3252
1.2800	1.0639	1.1901	2.7300	2.6420	1.8445
1.1000	1.0158	1.0789	3.2800	3.1053	3.2970
1.0000	1.0001	1.0003	4.6700	3.9214	4.4600
rmsd%	3.14	4.81		6.28	6.43

TABLE 4.2 Experimental and predicted activity coefficient for the binary system NBA + 2-ethyl-1-hexanol.

Activity coefficient NBA			Activity coefficient 2-ethyl-1-hexanol		
Experimental	UNIQUAC	NRTL	Experimental	UNIQUAC	NRTL
4.5500	4.5182	4.9910	1.0000	1.0000	1.0000
3.5700	3.3107	3.5758	1.1500	1.0164	1.0175
2.8200	2.5293	2.6902	1.2499	1.0657	1.0698
1.7720	1.7547	1.7248	1.4101	1.2772	1.2931
1.3500	1.2416	1.2729	1.6900	1.7005	1.7511
1.1220	1.0537	1.0630	2.8200	2.4901	2.6672
1.0000	1.0129	1.0156	3.5201	3.1131	3.4555
1.0000	1.0000	1.0000	4.3000	3.9704	4.6419
rmsd%	5.74	10.28		9.82	8.92

The experimental isobaric VLE data (T, x, y) for the binary systems (TBA + 2-ethyl-1-hexanol) and (NBA + 2-ethyl-1-hexanol) studied in this work are listed in Tables 4.3 and 4.4. The same experimental data together with calculated VLE data are graphically shown in Figures 4.3–4.6. The dashed line in theses diagrams were obtained from the UNIQUAC and NTRL equations. Antoine coefficients were used in calculating vapor pressures of TBA, 2-ethyl-1-hexanol, and NBA. Antoine coefficients (Table 4.5) were obtained by applying the equation:

$$\ln p_i^0 = A - \frac{B}{T - C} \qquad (3)$$

Vapor pressures, p_i^0 (mmHg), were fitted with the Antoine equation the parameters A, B and C, of pure component.

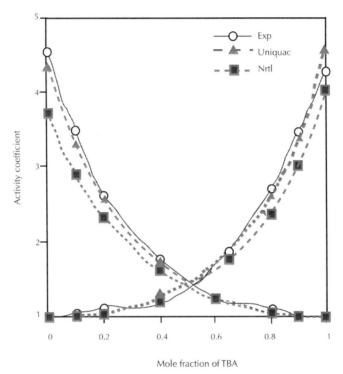

FIGURE 4.1 Activity coefficients in the system of TBA + 2-ethyl-1-hexanol.

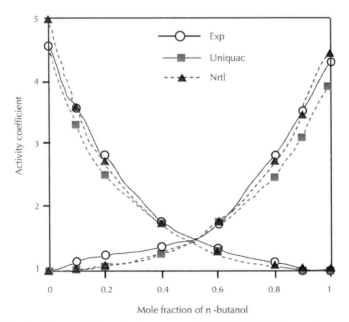

FIGURE 4.2 Activity coefficients in the system of NBA + 2-ethyl-1-hexanol.

Computer Calculations for Multicomponent Vapor–Liquid

TABLE 4.3 Vapor–liquid equilibria for *tert*-butanol (TBA) with 2-ethyl-1-hexanol.

T(K)	X₁	Y₁	T(K)	X₁	Y₁
459.7300	0.0000	0.0008	368.9010	0.3000	0.9730
445.5325	0.0050	0.2984	366.7212	0.4399	0.9797
436.1964	0.0100	0.4755	363.8160	0.5100	0.9819
428.8066	0.0150	0.5892	361.6598	0.6499	0.9858
422.7940	0.0199	0.6667	360.6300	0.7200	0.9877
388.2300	0.0899	0.9200	359.5657	0.7899	0.9899
380.5000	0.1599	0.9542	357.0900	0.8599	0.9925
372.1293	0.2299	0.9666	355.5700	0.9300	0.9957

TABLE 4.4 Vapor–liquid equilibria for *n*-butanol (NBA) with 2-ethyl-1-hexanol.

T(K)	X₁	Y₁	T(K)	X₁	Y₁
458.15	0.0000	0.0008	402.8900	0.3000	0.8703
452.6213	0.0049	0.2994	400.9500	0.3400	0.8845
444.5697	0.0112	0.3245	399.5000	0.4400	0.9110
441.3355	0.0200	0.3980	398.2821	0.5100	0.9234
418.6233	0.0900	0.7289	396.1976	0.6500	0.9235
409.6233	0.1685	0.8125	393.3350	0.8600	0.9567
405.6594	0.2300	0.9480	392.0000	0.9270	0.9768

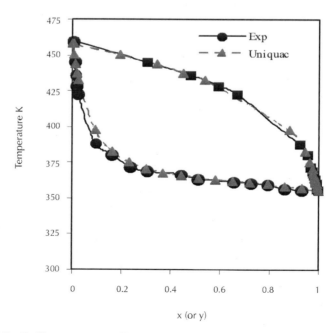

FIGURE 4.3 Boiling temperature diagram (T) for the system of TBA + 2-ethyl-1-hexanol.

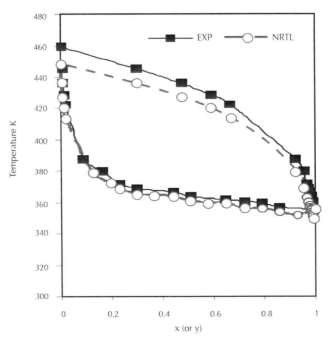

FIGURE 4.4 Boiling temperature diagram (T) for the system of TBA + 2ethyl-1-hexanol.

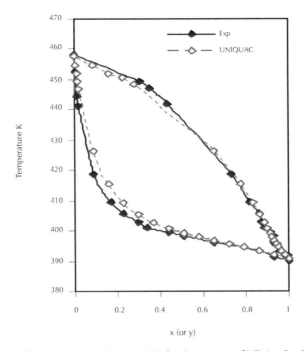

FIGURE 4.5 Boiling temperature diagram (T) for the system of NBA + 2-ethyl-1-hexanol.

Computer Calculations for Multicomponent Vapor–Liquid

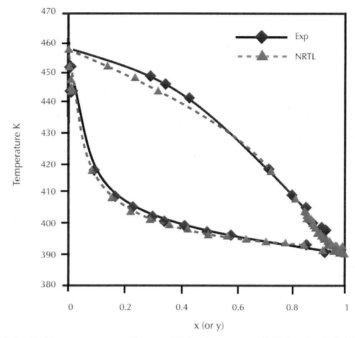

FIGURE 4.6 Boiling temperature diagram (T) for the system of NBA + 2-ethyl-1-hexanol.

TABLE 4.5 Antoine coefficients.

Component	A	B	C
n-butanol	17.2160	3137.02	-94.43
tert-butanol	16.8548	2658.29	-95.50
2-ethyl-1-hexanol	15.3614	2773.46	-140.00

In these T-x-y diagrams, it can be seen that the binary systems do not show azeotropic behavior as the boiling point of 2-ethyl-1-hexanol is quiet far from that those NBA and TBA. In contrast, the binary systems of water-TBA and water-NBA show azeotropic behavior. From the experimental results reported in our previous work, 2-ethyl-1-hexanol can be considered as a good solvent for recovering TBA and NBA from the aqueous solution.

TABLE 4.6 UNIQUAC and NRTL parameters the systems TBA + 2-ethyl-1-hexanol and NBA + 2-ethyl-1-hexanol.

I	j	UNIQUAC		NRTL	
		U_{12} (K)	U_{21} (K)	Δg_{12}	Δg_{21}
TBA	2-ethyl-1-hexanol	-55.4544	215.7780	511.769	480.0741
NBA	2-ethyl-1-hexanol	-78.3030	247.1739	384.4047	578.2420

Table 4.6 shows the calculated value of the UNIQUAC and NTRL binary interaction parameters for the mixture TBA + 2-ethyl-1-hexanol and NBA + 2-ethyl-1-hexanol using universal values for the UNIQUAC and NTRL parameters. The mixture non-randomness parameter α_{12} in the NTRL equation was fixed at 0.3. The values of r and q used in the UNIQUAC equation are presented in Table 4.7. The UNIQUAC structural parameters r and q were calculated from group contribution data that has been previously reported (Abrams and Prausnitz, 1975). The UNIQUAC and NTRL equations were optimized by minimizing the following objective function (OF):

$$OF = \sum_{j=1}^{np} \sum_{i=1}^{nc} \left[\frac{\gamma_{ij}^{exp} - \gamma_{ij}^{cal}}{\gamma_{ij}^{exp}} \right]^2 \qquad (4)$$

where γ_{ij} are the corresponding activity coefficients and np is the number of experimental data and nc is number components.

TABLE 4.7 The UNIQUAC structural parameters.

Components	R	q
TBA	3.45	3.05
NBA	3.45	3.05
2-ethyl-1-hexanol	6.15	5.02

The goodness of fit, between the observed and calculated mole fractions, was calculated in terms of the root mean square deviation (rmsd). The rmsd values were calculated according to the equation of percentage root mean square deviation (rmsd%):

$$RMSD\% = 100 \sqrt{\frac{1}{N} \sum_{i=1}^{N} \left(\frac{\gamma_i^{exp} - \gamma_i^{cal}}{\gamma_i^{exp}} \right)^2} \qquad (5)$$

where N is the number of data points, γ_i^{exp} indicates the experimental activity coefficient, γ_i^{cal} the calculated activity coefficient. The average (rmsd%) between the observed and calculated mol percents was 4.71% for TBA + 2-ethyl-1-hexanol and 7.78% for NBA + 2-ethyl-1-hexanol (predicted by UNIQUAC), whereas the average rmsd values 6.43 and 9.59% for systems TBA + 2-ethyl-1-hexanol and NBA + 2-ethyl-1-hexanol, respectively given by NRTL model. Values of rmsd given by these two models are listed in Table 4.8. From these values it can be concluded that the UNIQUAC model gives better prediction than NRTL for these binary systems.

TABLE 4.8 RMSD% values for the UNIQUAC and NRTL.

	NBA	TBA
UNIQUAC	5.74	3.14
	9.82	6.28
Average	7.78	4.71
NRTL	10.28	4.81
	8.92	8.05
Average	9.6	6.43

At liquid–liquid equilibrium, the composition of the two phases (refined phase and extracted phase) can be determined from the following equations

$$(\gamma_i x_i)_1 = (\gamma_i x_i)_2 \tag{6}$$

$$\Sigma x_{i1} = \Sigma x_{i2} = 1 \tag{7}$$

Here γ_{i1} and γ_{i2} are the corresponding activity coefficients of component i in phase 1 and 2, x_{i1}, and x_{i2} are the mole fraction of components i in the system and in phase 1 and 2 respectively. The interaction parameters between cyclohexane, methanol and ethyl benzene are used to estimate the activity coefficients from the UNIQUAC groups. Equations (1) and (2) are solved for the mole fraction (x) of component i in the two liquid phase. The UNIQUAC model (universal quasi–chemical model) is given by Abrams and prausnitz (Kojima and Tochigi, 1979) as

or

$$\ln \gamma_i = \ln \gamma_i^c + \ln \gamma_i^R \tag{8}$$

where

$$\frac{g^E}{RT} = \sum_{i=1}^{c} x_i \ln\left(\frac{\Phi_i}{x_i}\right) + \frac{z}{2}\sum_{i=1}^{c} q_i x_i \ln\left(\frac{\theta_i}{\Phi_i}\right) - \sum_{i=1}^{c} q_i x_i \ln\left(\sum_{j=1}^{c} \theta_j \tau_{ji}\right) \tag{9}$$

or

$$\ln \gamma_i = \ln \gamma_i^c + \ln \gamma_i^R \tag{10}$$

where

$$\ln \gamma_i^R = q_i \left[1 - \ln\left(\sum_{j=1}^{c} \theta_j \tau_{ji}\right) - \sum_{j=1}^{c} \left(\frac{\theta_j \tau_{ij}}{\sum_{k=1}^{c} \theta_k \tau_{kj}}\right) \right] \tag{11}$$

Here, γic is combinatorial parte of the activity coefficient, and γiR the residual part of the activity coefficient. The variable τij the adjustable parameter in the UNIQUAC equation and xi the equilibrium mole fraction of component i.

The parameter Φi (segment fraction) and θi (area fraction) are given by the following equation:

$$\Phi_i = \frac{x_i r_i}{\sum_{i=1}^{c} x_i r_i} \tag{12}$$

$$\theta_i = \frac{x_i r_i}{\sum_{i=1}^{c} x_i q_i} \tag{13}$$

$$\tau_{ij} = \exp\left(-\frac{(u_{ij} - u_{jj})}{RT}\right) \tag{14}$$

The parameter uij characterizes the interaction energy between compounds i and j and uij equals uji.

$$l_i = \left(\frac{z}{2}\right)(r_i - q_i) - (r_i - 1) \tag{15}$$

where z = 10, is lattice coordination number, ri the number of segments per molecule, and qi the relative surface area per molecule.

Figures 4.7 compare graphically the observed and calculated phase behavior (liquid-liquid equilibria data) for three ternary system, cyclohexane + methanol + benzene) at temperature of 288.2 K.

The liquid–liquid phase diagrams exhibit type 1 systems and as expected for these type of systems, the diagrams present plait point (where the two phases in equilibrium become experimentally miscible). Due to the variation of tie-line, the measuring of plait point is slightly difficult. Meanwhile, the value of the plait point is important and it is a necessary value to define the interval of solubility that present in components of a system. On other hand, this point can define the appropriate quantity of oxygenated compound that can be added to gasoline without phase separation. In view of the above, the plait points were determined using a graphic method (Abrams and Prausnitz, 1975). The values of the plait point for these systems are presented in Table 4.3.

As it can be observed from Figure 4.1, the ternary systems present a small region of partial miscibility that is limited by the plait point. This reveals that, methanol is totally miscible with the gasoline in a wide interval. As illustrated in Figure 4.1 and indicated in Table 4.4. It is evident that despite the representative compounds of the gasoline, the region of completely miscibility and also the plait point values are nearly the same and independent of the type of aromatic hydrocarbon. This provides an advantage as it can define the appropriate quantity of oxygenated compound (methanol) that can be added to the gasoline.

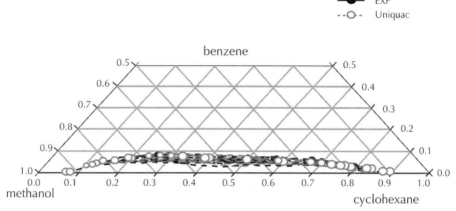

FIGURE 4.7 Experimental (—) and predicted UNIQUAC (---) LLE data at 288.2 K.

The UNIQUAC model was successfully used to correlate the experimental LLE data. As it can be seen from Figure 4.1, the predicted tie lines (dashed lines) are in good agreement with the experimental data (solid lines). In other words, the UNIQUAC equations adequately fit the experimental data for this multi-component system. The optimum UNIQUAC interaction parameters uij between cyclohexane, methanol, and benzene were determined using the observed liquid–liquid data, where the interaction parameters describe the interaction energy between molecules i and j or between each pair of compounds. Table 4.4 show the calculated value of the UNIQUAC binary interaction parameters for the mixture methanol + benzene using universal values for the UNIQUAC structural parameters. The equilibrium model was optimized using an OF, which was developed by Sørensen (1980).

Moreover, the OF obtained by minimizing the square of the difference between the mole fractions calculated by UNIQUAC model and the experimental data. Furthermore, the UNIQUAC structural parameters r and q were carried out from group contribution data that has been previously reported (Abrams and Prausnitz, 1975; Alkandary et al., 2001; Arce et al., 1995; Fabries et al., 1977; Fernandez-Torres et al., 1999; Garcia-Flores et al., 2001; Ghanadzadeh, 1993b; Ghanadzadeh and Ghanadzadeh, 2003b; Helinger and Sandler, 1995; Higashiuchi et al., 1990; Pesche and Sandler, 1995; Prausnitz et al., 1980; Sorensen, 1980; Velo Garcia, 1992; Wisniewska-Goclowska and Malanowski, 2001). The values of r and q used in the UNIQUAC equation are presented in Table 4.4. The goodness of fit, between the observed and calculated mole fractions, was calculated in terms rmsd (Artigas et al., 1997). The rmsd values were calculated according to the equation of rmsd% where n is the number of tie-lines, xexp indicates the experimental mole fraction, xcalc is the calculated mole fraction, and the subscript i indexes components, j phases and $k = 1,2,...n$

(tie-lines). The average (rmsd%) between the observed and calculated mole percents with a reasonable error was 3.57% cyclohexane + methanol + benzene (see Table 4.4). The percentage of relative error between the experimental and predicted values of the plait point for these systems have been also compiled in Table 4.4.

In Figure 4.2 the separation factor (S) of methanol as a function of the mole fraction of methanol in MCH phase, indicate that the factor of separation increases, as the methanol solubility increases in cyclohexane. The experimental result shows that the existence of aromatic compound (benzene) in gasoline increases the solubility of methanol in cyclohexane.

4.4 CONCLUSION

Isobaric VLE data at 101.3 kPa were measured for two binary mixtures, that is TBA + 2-ethyl-1-hexanol and NBA + 2-ethyl-1-hexanol. The UNIQUAC and NRTL models were used to correlate the experimental activity coefficients and VLE data. Deviation from ideal solution behavior was observed for both binary systems, whereas none of them showed azeotropic behavior.

The UNIQUAC model fit the experimental data for the systems of TBA + 2-ethyl-1-hexanol and NBA + 2-ethyl-1-hexanol with an average RMSD of 4.71% and 6.43% respectively. The NRTL model gave higher values for the systems, that is 7.78% and 9.59% respectively. From these values it can be concluded that the NRTL model gives poor predictions, whereas the UNIQUAC model adequately fit the experimental VLE data for these binary systems.

KEYWORDS

- **Gas chromatography**
- **Liquid liquid equilibria**
- **Multi-component system**
- **Non-random two-liquid model**
- **Objective function**
- **Universal quasi-chemical model**
- **Vapor–liquid equilibrium**

REFERENCES

Abrams, D. S. and Prausnitz, J. M. (1975). *AICHE J.* **21**, 116.

Alkandary, J. A., Aljimaz, A. S., Fandary, M. S., and Fahim, M. A. (2001). *Fluid Phase Equilibria* **187–188**, 131.

Arce, A., Blanco, A., Martinez-Ageitos, J., and Vidal, I. (1995). *Fluid Phase Equilibria* **109**, 291.

Artigas, H., Lafuente, C., Lopez, M. C., Royo, F. M., and Urieta, J. S. (1997). *Fluid Phase Equilib.* **134**, 163.

Artigas, H., Lafuente, C., Martin, S., Minones, J., Jr., and Royo, F. M. (2001). *Fluid Phase Equilib.* **192**, 49.

Darwish, N. A. and Al-Anber, Z. A. (1997). *Fluid Phase Equilib.* **131**, 287.

Fabries, J. F., Gustin, J. L., and Renon, H. (1977). *Chem. Eng. Data* **22**, 303–308.

Fernandez-Torres, M. J., Gomis-Yagues, V., Ramos-Nofuentes, M., and Ruiz-Bevia, F. (1999). *Fluid Phase Equilibria* **164**, 267.

Garcia-Flores, B. E., Galicia-Aguilar, G., Eustaquio-Rincon, R., and Trejo, A. (2001). *Fluid Phase Equilibria* **185**, 275.

Ghannadzadeh, H. (1993a). *Election of solvent selective for the extraction in phase liquid alcohols C4 (ABE) starting from biomass*. Ph.D. Thesis, Polytechnic Universitat of catalunya Barcelona, Spain.

Ghanadzadeh, H. (1993b). *Eleccion de disolventes selectivos para la extraccion en fase liquida de alcoholes C4 (ABE) a partir de biomasa*. Ph.D. Thesis, Universitat Politecnica de catalunya Barcelona, Spain.

Ghanadzadeh, H. and Ghanadzadeh, A. (2002a). *Fluid Phase Equilib.* **202**, 337.

Ghanadzadeh, H. and Ghanadzadeh, A. (2002b). *Fluid Phase Equilibria* **202**, 339.

Ghanadzadeh, H. and Ghanadzadeh, A. (2003a). *J. Chem. Thermodynamics.* in press.

Ghanadzadeh, H. and Ghanadzadeh, A. (2003b). *J. Chem. Thermodyn* **35**, 1393–1401.

Helinger, S. and Sandler, S. I. (1995). *J. Chem. Eng. data* **40**, 321.

Higashiuchi, H., Sakuragi, Y., Arai, Y., and Nagatani, M. (1990). *Fluid Phase Equilibria* **58**, 147.

Hirata, M., Ohe, S., and Nagakama, K. (1975). *Computer-Aided Data Book of vapor-liquid equilibria*. Elsevier, New York.

Kojima, K. and Tochigi, K. (1979). *Prediction of vapor-liquid equilibria by the ASOG Method*. Elsevier, Tokyo.

Pesche, N. and Sandler, S. I. (1995). *J. Chem. Eng. Data* **40**, 315.

Prausnitz, J. M., Anderson, T. F., Grens, E. A., Eckert, C. A., Hsien, R., and Oconnell, J. P. (1980). *Computer Calculations for Multicomponent Vapor-Liquid and Liquid-Liquid Equilibria*. Prentice-Hall, Inc, Englewood.

Quitzsch, K., Koop, R., Renker, W., and Geiseler, G. (1968). *Z. Phys. Chem.* **237**, 265.

Renon, H. and Prausnitz, J. M. (1968). *AIChE J.* **14**, 135.

Rodriguez, A., Canosa, J., Domenguez, A., and Tojo, J. (2002). *Fluid Phase Equilib.* **198**, 95.

Seo, J., Canosa, J., Lee, J., and Kim, H. (2000). *Fluid Phase Equilib.* **172**, 211.

Sorensen, J. M. (1980). Correlation of liquid–liquid equilibrium data. Ph.D. Thesis,Technical University of Denmark, Lyngby, Denmark.

Suska, J. E., Holub, R., Vonka, P., and Pick, J. Coll. (1970). *Czech. Chem. Commun.* **35**, 385.

Velo Garcia, E. (1992). Cinetica equilibria y transport de materia en la hidratacion catalitica directa de isobuteno a tert-butanol. Ph.D. Thesis, Universitat Politecnica de catalunya Barcelona, Spain.

Walas, S. M. (1985). *Phase Equilibria in Chemical Engineering*, Butterworths, London.

Wisniewska-goclowska, B. and Malanowski, S. K. (2001). *Fluid Phase Equilibria* **180**, 103.

5 Densities and Refractive Indices of the Binary Systems

CONTENTS

5.1 Introduction 43
5.2 Experimental 44
    5.2.1 Chemical 44
5.3 Results and Discussion 44
5.4 Conclusion 47
Keywords 47
References 47

5.1 INTRODUCTION

There is an increased interest in the thermodynamic behavior of liquid mixtures. In this way, a considerable amount of work has been carried out in the last years on the experimental determination of the density and refractive index of binary and ternary mixtures, and the corresponding excess molar volume and change in refractive index (Al-Jimaz et al., 2005; Francesconi et al., 2000; Joshi and Aminabhavi, 1990; Orge et al., 2000; Viswanathan et al., 2000).

The present work is part of our systematic studies on thermodynamic properties for mixtures of great interest in industry. Knowledge of these mixing properties has relevance in both theoretical and applied areas of research because such results are useful in design and simulation process. In a previous work (Ottani and Comelli, 2005), we have studied the vapor–liquid equilibrium of the binary systems ethanol alcohol, and toluene, because of the importance of the ethanol as additive of reformulated gasoline. The present work reports densities and refractive indices for the binary systems mentioned previously, the binary system ethanol + toluene at 298.15 K, and the corresponding excess molar volumes and changes of refractive indices.

Experimental values of excess molar volume and change of refractive index of binary mixtures were correlated by Redlich–Kister equation (Ottani and Comelli, 2005). The binary contribution was used to correlate the experimental ternary results with the Cibulka equation (Al-Jimaz et al., 2004). Several methods to predict density and refractive index of mixtures were applied to test their validity in the systems studied.

Density measurements for ethanol were carried out at several temperatures to obtain parameters in some predictive density methods.

5.2 EXPERIMENTAL

5.2.1 Chemical

The chemicals ethanol, toluene 99.5% were supplied by Aldrich. The purity of all chemicals was checked by gas chromatography (GC) and the results of these analysis showed that the impurities did not exceed 0.2 mass%. The pure components were degassed ultrasonically. These reagents were used without further purification. Further verification was realized by ascertaining the constancy of the values of density and refractive index for every component at 298.15 K, which were reasonably in accordance with values found in the literature. (Grunberg and Nissan, 1949; Heric and Brewer, 1967; McAllister, 1960; Redlich and Kister, 1948; Tasiæ et al., 1992).

The samples were prepared using a Mettler AE 200 balance with an accuracy of 0.0001 g, covering the whole composition range of the binary and ternary mixtures. Precautions were taken such as using samples recently prepared and reducing the vapor space in the vessels to a minimum, in order to avoid preferential evaporation during manipulation and subsequent composition errors. The accuracy in the determination of the mole fraction of the measured samples was also 0.0001.

The densities of pure components and their mixtures were measured with an Anton Paar DMA 55 densimeter, and the refractive indices with an Abbe refractometer Type 3T., with accuracies of 0.01 kg my3, and 0.0002, respectively. Each apparatus was matched to a Julabo circulator with proportional temperature control and an automatic drift correction system that kept the samples at desired temperature usually 298.15 K with accuracy of 0.01 K. The densimeter was calibrated periodically with distilled water and dry air.

5.3 RESULTS AND DISCUSSION

Densities, r, refractive indices, n, excess molar volumes, VE, and changes in refractive indices, Ddn, of binary and ternary mixtures are shown in Table 5.1. In these tables, x is the molar fraction DI of component i in the mixture.

The excess molar volumes and the changes in refractive indices on mixing for binary mixtures were calculated from equations (1) and (2), respectively:

$$V^E = \sum_{i=1}^{N} x_i M_i \left(\rho^{-1} - \rho_i^{-1} \right) \qquad (1)$$

$$\delta n_d = n_d - \sum_{i=1}^{N} x_i n_{di} \qquad (2)$$

where M is the molecular weight of component i in the mixture, ρ_i and n_d are the properties of the pure components and N is the number of components in the mixture.

The derived binary excess properties were correlated with the Redlich–Kister equation, according to the expression:

Densities and Refractive Indices of the Binary Systems

$$\Delta E = x_i x_j \sum_{t=0}^{m} A_t (x_i - x_j)^2 \qquad (3)$$

In this equation, ΔE is the excess property, A_t is a parameter and m is the degree of the polynomial expansion fitting equation (3). The corresponding standard deviations obtained are given in Table 5.2.

TABLE 5.1 Densities, ρ, refractive indices, nD, excess molar volumes, VE, and changes of refractive indices on mixing δnD for the binary system ethanol (1) + toluene(2) at 298.15 K.

X1xxx0000	Density	VE	nD	δn
0.000049	861.0000	0.0000	1.4901	0.0000
0.0549	857.3244	0.1129	1.4950	0.0001
0.0726	855.8900	0.1460	1.4931	0.0002
0.1326	851.0081	0.2491	1.4865	0.0004
0.1912	846.2404	0.3352	1.4801	0.0004
0.3086	836.6967	0.4645	1.4672	0.0004
0.3850	830.4864	0.5176	1.4588	0.0003
0.4116	828.3267	0.5304	1.4559	0.0003
0.4669	823.8279	0.5475	1.4498	0.0003
0.5149	819.9240	0.5520	1.4446	0.0001
0.5935	813.5344	0.5386	1.4359	0.0001
0.6144	811.8392	0.5307	1.4336	0.0000
0.6624	807.9354	0.5056	1.4284	−0.0001
0.7264	802.7357	0.4572	1.4214	−0.0002
0.8032	796.4846	0.3764	1.4129	−0.0002
0.8437	793.1920	0.3238	1.4085	−0.0002
0.8871	789.6631	0.2600	1.4037	−0.0002
0.9287	786.2808	0.1914	1.3992	−0.0002
0.9706	782.8741	0.1149	1.3946	−0.0001
0.0000	779.5000	0.0000	1.3900	0.0000

TABLE 5.2 Parameters of equation (3) standard deviations, s, at 298.15 K.

	A_0	A_1	A_2	A_3
P	1657.5	−4830.7	16706	9215
VE	1.3022	−1.4756	6.1560	1.8956
n_d	2.912	8.607	29.493	16.794
δn_D	0.0008	0.0031	−0.0007	−0.0003

As can be seen in Figures 5.1 and 5.2, all the binary systems show positive values of the excess molar volume over the whole range of composition, with a maximum at nearly equimolecular composition. From comparison of different ethanol + toluene binary systems Figure 5.1.

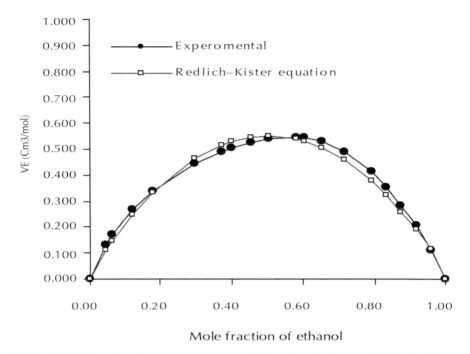

FIGURE 5.1 Variation of excess molar volume (VE) with mole fraction (x1) for the binary systems at T) 298.15 K:,ethanol + toluene.

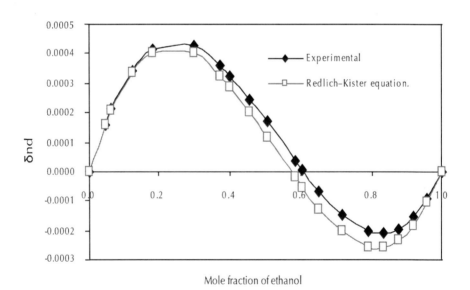

FIGURE 5.2 Changes in refractive indices, δnD, as a function of composition ethanol.

5.4 CONCLUSION

Densities and excess molar volumes for the binary liquid mixtures of ethanol (1) + 2ethyl-1-hexanol (2) were measured at the temperature of 298.15 K and atmospheric pressure over the whole range of compositions. Excess molar volumes are positive, are significant. The geometrical models used to predict ternary properties from binary contributions produce, in general, average absolute deviations lower than 33% for molar excess volume.

KEYWORDS

- Binary systems
- Density and refractive index
- Liquid mixtures
- Molar volume
- Redlich–Kister equation
- Ternary mixtures

REFERENCES

Al-Jimaz, A. S., Al-Kandary, J. A., and Abdu-Latif, A. M. (2004). Viscosities and densities for binary mixtures of phenetole with 1-pentanol, 1-hexanol, 1-heptanol, 1-octanol, 1-nonanol, and 1-decanol at different temperatures. *Fluid Phase Equilib.* 218, 247–260.

Al-Jimaz, A. S., Al-Kandary, J. A., Abdu-Latif, A. M., and Al-Zanki, A. M. (2005). Physical properties of anisole + *n*-alkanes at (temperatures between 293.15 and 303.15 K). *J. Chem. Thermodyn.* 37, 631–642.

Francesconi, R., Comelli, F., and Castellari, C. (2000). Excess molar enthalpies and excess molar volumes of binary mixtures containing dialkyl Carbonates + Anisole or Phenetole at (288.15 and 313.15) K. *J. Chem. Eng. Data* 45, 544–548.

Grunberg, L. and Nissan, A. H. (1949). Mixture law for viscosity. *Nature* 164, 799–800.

Heric, E. L. and Brewer, J. G. (1967). Viscosity of some binary liquid nonelectrolyte mixtures. *J. Chem. Eng. Data* 12, 574–583.

Joshi, S. S. and Aminabhavi, T. M. (1990). Densities and viscosities of binary liquid mixtures of anisole with methanol and benzene. *J. Chem. Eng. Data* 35, 187–189.

McAllister, R. A. (1960). The viscosity of liquid mixtures. *AIChE J.* 6, 427–431.

Orge, B., Marino, G., Iglesias, M., Tojo, J., and Pin˜eiro, M. M. (2000). Thermodynamics of (anisole + benzene, or toluene, or *n*-hexane or cyclohexane or 1-butanol, or 1-pentanol) at 298.15 K. *J. Chem. Thermodyn.* 32, 617–629.

Ottani, S. and Comelli, F. (2005). Excess enthalpies of biary mixtures containing poly(propylene glycols) + benzyl alcohol, or + *m*-cresol, or + anisole at 308.15 K and at atmospheric pressure. *Thermochim. Acta* 430, 123–128.

Redlich, O. and Kister, A. T. (1948). Algebric representation of thermodynamic properties and classification of solutions. *Ind. Eng. Chem.* 40, 345–348.

Tasiæ, A., Djordjevic, B., and Grozdaniæ, D. (1992). Use of mixing rules in predicting refractive indices and specific refractivity's for some binary liquid mixtures. *J. Chem. Eng. Data* 37, 310–331.

Viswanathan, S., Rao, M. A., and Prasad, D. H. L. (2000). Densities and viscosities of binary liquid mixtures of anisole or methyl *tert*-butyl ether with benzene, chlorobezene, benzonitrial, and nitrobenzene. *J. Chem. Eng. Data* 45, 764–770.

6 Potential Applications of Artificial Neural Networks to Thermodynamics

CONTENTS

6.1 Introduction 49
6.2 Group Method of Data Handling (GMDH) 50
    6.2.1 GMDH Algorithm 50
    6.2.2 Prediction of LLE using the GMDH-Type Network 53
6.3 Conclusion 58
Keywords 58
References 59

6.1 INTRODUCTION

The importance of the availability of precise liquid–liquid equilibrium (LLE) data in rational design of many chemical processes and separation operations, have been the subject of much research in recent years. A large amount of investigation has been carried out on the LLE measurements, in order to understand and provide further information about the phase behavior of such systems. Usually, the equilibrium data presented are correlated using thermodynamic methods. The thermodynamic models have been successfully applied for the correlation of several LLE systems but these conventional methods for LLE data prediction of complex systems are tedious. Recently, to avoid these limitations, new prediction methods were developed by using artificial neural network (ANN). The ANNs are non linear and highly flexible models that have been successfully used in many fields to model complex non-linear relationships. Hence they offer potential to overcome the limitations of the traditional thermodynamic models and polynomial correlation methods for the complicated systems, especially in estimating the LLE and vapor–liquid equibilirum (VLE) (Ganguly, 2003; Guimaraes and McGreavy, 1995; Mohanty, 2005; Sharma et al., 1999; Urata et al., 2002). The ANNs may be viewed as the universal approximators but the main disadvantage of them is that the detected dependencies are hidden within the neural network (NN) structure (Aksenova et al., 2005). Conversely, Group Method of Data Handling (GMDH) (Ivakhnenko, 1971) is aimed to identify the functional structure of a model hidden in the empirical data. The main idea of GMDH is the use of feed-forward networks based on short-term polynomial transfer function whose coefficients are obtained using regression technique combined with the emulation of the self-organizing activity for the NN structural learning (Farlow, 1984). The GMDH was developed for complex systems modeling, prediction, identification and approximation of multivariate processes, diagnostics, pattern recognition, and clusterization of data sample. It was proved, that for inaccurate, noisy or small data can be found best optimal simplified model, accuracy of which is higher and structure is simpler than structure of usual full physical model. In this work, to avoid the limitations of ANNs, a LLE prediction method was developed by using GMDH algorithm. The aim of this proposed method is to predict LLE data of a quaternary system (Goncalves et al., 2002), Corn oil + oleic acid + ethanol + water, using GMDH algorithm. Using existing data in (Goncalves et al., 2002), the proposed network was trained and the trained network used to predicting of LLE data in oil phase and alcohol phase. Then, the predicted data of the proposed model compared with the experimental data. Also mean deviations obtained by NRTL, UNIQUAC, and proposed model have been compared. The phase diagrams for the studied quaternary system including both the experimental and predicted tie lines are presented.

6.2 GROUP METHOD OF DATA HANDLING (GMDH)

The GMDH is a combinatorial multi-layer algorithm in which a network of layers and nodes is generated using a number of inputs from the data stream being evaluated. The GMDH was first proposed by Alexy G. Ivakhnenko (Ivakhnenko, 1971). The GMDH network topology has been traditionally determined using a layer by layer pruning process based on a pre-selected criterion of what constitutes the best nodes at each level. The goal is to obtain a mathematical model of the object under study. The GMDH creates adaptively models from data in form of networks of optimized transfer functions in a repetitive generation of layers of alternative models of growing complexity and corresponding model validation and fitness selection until an optimal complex model which is not too simple and not too complex has been created. Neither, the number of neurons and the number of layers in the network, nor the actual behavior of each created neuron are predefined. All these are adjusted during the process of self-organization by the process itself. As a result, an explicit analytical model representing relevant relationships between input and output variables is available immediately after modeling. This model contains the extracted knowledge applicable for interpretation, prediction, classification or diagnosis problems (Onwubolu, 2007).

6.2.1 GMDH Algorithm

The traditional GMDH method (Farlow, 1984; Ivakhnenko, 1971) is based on an underlying assumption that the data can be modeled by using an approximation of the Volterra Series or Kolmorgorov–Gabor polynomial (Madala and Ivakhnenko, 1994) as shown in equation (1).

$$y = a_0 + \sum_{i=1}^{m} a_i x_i + \sum_{i=1}^{m}\sum_{j=1}^{m} a_{ij} x_i x_j + \sum_{i=1}^{m}\sum_{j=1}^{m}\sum_{k=1}^{m} a_{ijk} x_i x_j x_k ... \qquad (1)$$

where x_i, x_j, x_k are the inputs, y the output and a_0, a_i, a_{ij}, a_{ijk} are the coefficients of the polynomial functional node.

A GMDH network can be represented as a set of neurons in which different pairs of them in each layer are connected through a quadratic polynomial and thus produce new neurons in the next layer (Nariman-Zadeh et al., 2002). In the classical GMDH algorithm, all combinations of the inputs are generated and sent into the first layer of the network. The outputs from this layer are then classified and selected for input into the next layer with all combinations of the selected outputs being sent into layer 2. This process is continued as long as each subsequent layer (n+1) produces a better result than layer (n). When layer (n+1) is found to not be as good as layer (n), the process is halted. The formal definition of the problem is to find a function $\hat{f}$ so that can be approximately used instead of actual one, f, in order to predict output $\hat{y}$ for a given input vector $X = (x_1, x_2, x_3, \ldots, x_n)$ as close as possible to its actual output y. Therefore, given M observation of multi-input–single-output data pairs (*training data set*) so that

$$y_i = f(x_{i1}, x_{i2}, x_{i3}, \ldots, x_{in}), \quad i=1, 2, \ldots, M \tag{2}$$

It is possible to train a GMDH-type network to predict the output values $\hat{y}$ using training data that is

$$\hat{y}_i = \hat{f}(x_{i1}, x_{i2}, x_{i3}, \ldots, x_{in}), \quad i=1, 2, \ldots, M \tag{3}$$

This equation is tested for fit by determining the mean square error of the predicted $\hat{y}$ and actual y values as shown in equation (4) using the set of *testing* data.

$$\sum_{i=1}^{M}(\hat{y}_i - y_i)^2 \to \min. \tag{4}$$

General connection between inputs and output variables can be expressed by equation (1). For most application the quadratic form of only two variables is used in the form to predict the output $\hat{y}$.

$$\hat{y} = G(x_i, x_j) = a_0 + a_1 x_{i_n} + a_2 x_{j_n} + a_3 x_{i_n} x_{j_n} + a_4 x_{i_n}^2 + a_5 x_{j_n}^2 \tag{5}$$

A typical feed-forward GMDH-type network is shown in Figure 6.2. The coefficients a_i in equation (5) are calculated using regression techniques (Farlow, 1984; Ivakhnenko, 1971) so that the difference between actual output, y, and the calculated one, $\hat{y}$, for each pair of x_i, x_j as input variables is minimized. Indeed, it can be seen that a tree of polynomials is constructed using the quadratic form given in equation (5) whose coefficients are obtained in a least-squares sense. In this way, the coefficients of each quadratic function G_i are obtained to optimally fit the output in the whole set of input–output data pair that is

$$r^2 = \frac{\sum_{i=1}^{M}(y_i - G_i(\ ))^2}{\sum_{i=1}^{M} y_i^2} \to \min. \tag{6}$$

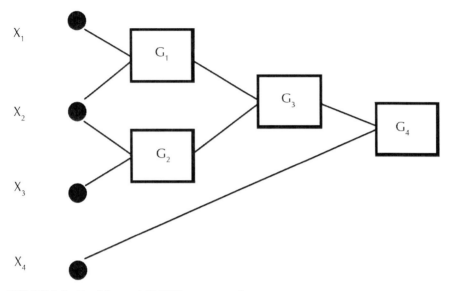

FIGURE 6.1 Feed-forward GMDH-type network.

In the basic form of the GMDH algorithm, all the possibilities of two independent variables out of total n input variables are taken in order to construct the regression polynomial in the form of equation (5) that best fits the dependent observations (y_i, $i = 1, 2, \ldots, M$) in a least-squares sense. Consequently,

$$\binom{n}{2} = \frac{n(n-1)}{2}$$

neurons will be built up in the second layer of the feed-forward network from the observations $\{(y_i, x_{ip}, x_{iq}), (i = 1, 2, \ldots, M)\}$ for different $p, q \in \{1, 2, \ldots, M\}$ (Farlow, 1984). In other words, it is now possible to construct M data triples $\{(y_i, x_{ip}, x_{iq}), (i = 1, 2, \ldots, M)\}$ from observation using such $p, q \in \{1, 2, \ldots, M\}$ in the form

$$\begin{bmatrix} x_{1p} & x_{1q} & \vdots & y_1 \\ x_{2p} & x_{2q} & \vdots & y_2 \\ \ldots & \ldots & \ldots & \ldots \\ x_{Mp} & x_{Mq} & \vdots & y_M \end{bmatrix}$$

Using the quadratic sub-expression in the form of equation (5) for each row of M data triples, the following matrix equation can be readily obtained as

$$Aa = Y \qquad (7)$$

where a is the vector of unknown coefficients of the quadratic polynomial in equation (5):

$$a = \{a_0, a_1, a_2, a_3, a_4, a_5\} \tag{8}$$

and

$$Y = \{y_1, y_2, y_3, \ldots, y_M\}^T \tag{9}$$

is the vector of output's value from observation. It can be readily seen that

$$A = \begin{bmatrix} 1 & x_{1p} & x_{1q} & x_{1p}x_{1q} & x_{1p}^2 & x_{1q}^2 \\ 1 & x_{2p} & x_{2q} & x_{2p}x_{2q} & x_{2p}^2 & x_{2q}^2 \\ \ldots & \ldots & \ldots & \ldots & \ldots & \ldots \\ 1 & x_{Mp} & x_{Mq} & x_{Mp}x_{Mq} & x_{Mp}^2 & x_{Mq}^2 \end{bmatrix} \tag{10}$$

The least-squares technique from multiple-regression analysis leads to the solution of the normal equations in the form of

$$a = \left(A^T A\right)^{-1} A^T Y \tag{11}$$

which determines the vector of the best coefficients of the quadratic equation (5) for the whole set of M data triples.

6.2.2 Prediction of LLE using the GMDH-type Network

The proposed model is a feed-forward GMDH-type network and has constructed using experimental data set from ref. (Goncalves et al., 2002). This data set is constituted of 25 points in four different concentrations of water in solvent. In Table 6.1, the overall experimental compositions of the mixtures and in Table 6.2, experimental mass fractions of the components in alcohol and oil phase are shown. The data set is divided in two parts, 80% used as *training* and 20% used as *testing* data. Each point in *training* and *test* data is constituted of 13 values. The 4 mass fractions in overall compositions and water concentration in solvent are normalized and used as inputs of GMDH-type network ($X_1, \ldots, X_5$) and other 8 values are used as desired outputs of network, 4 mass fractions in alcohol phase ($Y_1, \ldots, Y_4$) and 4 mass fractions in oil phase ($Z_1, \ldots, Z_4$). After applying the data set to the network, GMDH-type network eight polynomial equations are obtained that can be used to predicting of mass fractions in alcohol and oil phase. For example, the prediction equations of mass fraction of the acid in alcohol and oil phase are:

$$Y_2 = 1.15753 X_3 - 7.53321 X_1 X_2 X_3$$

$$Z_2 = 1.96267 X_3^2 + 1.50010 X_2 X_3 + 15.4356 X_2 X_3 X_5$$

where X_1 is the water concentration in solvent and X_2, X_3 and X_5 are the normalized mass fraction of oleic acid, ethanol and water in overall composition, respectively. The network topology of this part of GMDH model is shown in Figure 6.2.

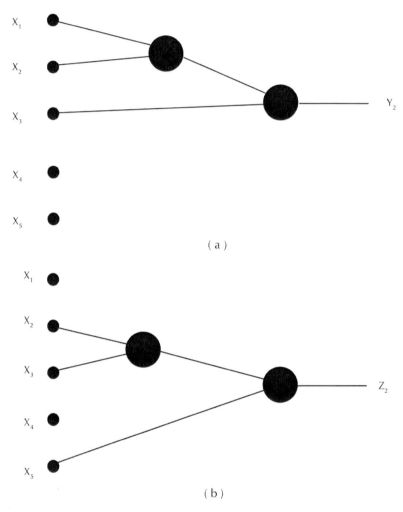

FIGURE 6.2 The GMDH-type network topology for mass fraction of acid in (a) alcohol phase (b) oil phase.

We used GMDH model to calculate the mass fractions of the components in alcohol and oil phase. The calculated values are presented in Table 6.2. Figure 6.3 and 6.4 show the experimental points and predicted tie lines from GMDH model for the systems corn oil/oleic acid/5% aqueous ethanol and corn oil/oleic acid/8% aqueous ethanol. The equilibrium diagrams were plotted in triangular coordinates. For representing the pseudo ternary systems in triangular coordinates, ethanol + water was admitted as a mixed solvent (Goncalves et al., 2002). These Figures indicate that GMDH model provided a good estimation in both phases.

TABLE 6.1 The overall composition of LLE data for the system + solvent (1) + ethanol (3) + water (4) at 298.15 K.

Water conc. in solvent	Overall composition			
	$100w_1$	$100w_2$	$100w_3$	$100w_4$
5 wt %	47.98	0.00	49.40	2.63
	47.21	2.53	47.72	2.54
	43.46	4.91	49.02	2.61
	39.25	9.87	48.32	2.57
	35.65	14.52	47.32	2.51
	29.85	19.99	47.62	2.53
8 wt %	49.97	0.00	46.03	4.00
	44.97	5.39	45.67	3.97
	39.78	9.81	46.38	4.03
	35.49	14.59	45.93	3.99
	30.99	19.77	45.30	3.94
12 wt %	50.07	0.00	43.94	5.99
	47.94	2.40	43.70	5.96
	45.85	4.92	43.32	5.91
	41.49	9.65	43.26	5.90
	34.15	14.79	44.93	6.13
	30.04	19.99	43.97	5.99
	24.59	25.06	44.30	6.04
18 wt %	50.35	0.00	40.72	8.94
	48.27	2.42	40.44	8.88
	44.10	4.91	41.81	9.18
	39.94	9.80	41.22	9.05
	34.70	15.08	41.18	9.04
	29.66	20.15	41.16	9.03
	25.22	24.89	40.91	8.97

Figure 6.5 presents the fatty acid distribution between the phases and Figure 6.6 shows the experimental and estimated solvent selectivities.

The distribution coefficient and solvent selectivity can be calculated by equations (12) and (13) respectively

$$k_i = \frac{w_i^{II}}{w_i^{I}} \tag{12}$$

$$S = \frac{k_2}{k_1} \tag{13}$$

TABLE 2 The experimental and predicted liquid–liquid equilibrium data for the studied system in alcohol and solvent.

Water conc. in solvent	Water phase (II)									Cyclohexane I phase (I)							
	$100w_1$		$100w_2$		$100w_3$		$100w_4$		$100w_1$		$100w_2$		$100w_3$		$100w_4$	
	Exp.	Calc.	Exp.	Calc.	Exp.	Calc.	Exp.	Calc.	Exp.	Calc.	Exp.	Calc.	Exp.	Calc.	Exp.	Calc.
5 wt %	1.61	1.42	0.00	0.00	92.39	92.79	5.99	5.79	91.63	91.99	0.00	0.00	8.07	7.73	0.30	0.28
	2.33	2.19	2.40	2.54	89.93	89.40	5.34	5.87	87.79	84.64	2.24	2.38	9.65	12.53	0.33	0.45
	1.61	2.08	5.11	4.89	87.91	87.36	5.37	5.66	84.23	84.67	4.64	4.53	10.74	10.45	0.39	0.34
	4.33	4.21	10.26	10.01	80.39	80.50	5.03	5.28	75.20	75.30	9.35	9.24	14.89	14.94	0.56	0.53
	7.35	7.42	15.11	14.83	73.06	73.09	4.48	4.66	65.77	65.18	13.87	13.96	19.70	20.08	0.67	0.77
	16.72	16.50	20.25	20.76	59.17	59.09	3.86	3.65	50.11	50.95	19.29	18.99	28.51	28.03	2.09	2.03
8 wt %	0.66	0.77	0.00	0.00	88.38	88.24	10.96	10.98	93.76	93.20	0.00	0.00	5.64	6.16	0.60	0.63
	1.34	1.18	4.54	4.77	83.36	83.26	10.76	10.80	85.34	85.15	5.64	5.64	8.36	8.47	0.66	0.73
	1.71	1.44	8.73	8.91	79.45	79.29	10.11	10.37	77.96	79.54	10.39	10.12	10.88	9.60	0.76	0.73
	2.57	2.85	13.82	13.71	73.76	73.61	9.86	9.83	69.63	70.21	15.34	15.16	13.91	13.59	1.11	1.04
	5.14	5.40	19.33	19.24	66.49	66.40	9.03	8.95	58.97	58.71	20.97	20.85	18.40	18.80	1.66	1.63
12 wt %	0.44	0.66	0.00	0.00	85.59	85.65	13.97	13.69	94.57	94.85	0.00	0.00	5.10	4.81	0.34	0.34
	0.67	0.79	1.81	1.74	83.73	83.73	13.80	13.74	90.56	90.39	2.71	2.88	6.08	6.21	0.65	0.52
	0.82	0.85	3.80	3.68	81.62	81.74	13.76	13.73	86.09	85.83	5.65	5.86	7.59	7.63	0.66	0.68
	1.21	1.11	7.86	7.55	77.73	78.05	13.21	13.28	78.14	78.37	10.97	11.42	10.13	9.46	0.77	0.74
	2.03	1.75	12.99	12.66	72.49	72.87	12.49	12.72	69.08	69.10	16.54	16.58	13.37	13.33	1.01	0.99
	3.98	3.43	18.34	17.96	66.19	66.78	11.48	11.83	59.72	58.39	21.67	22.46	17.08	17.48	1.53	1.67
	8.31	9.19	24.04	25.00	57.41	54.65	10.24	11.17	48.27	43.92	26.35	26.28	22.83	26.34	2.55	3.46
18 wt %	0.20	0.07	0.00	0.00	79.52	79.27	20.28	20.66	95.71	95.84	0.00	0.00	3.68	3.51	0.61	0.65
	0.19	0.19	1.43	1.23	77.88	77.65	20.49	20.92	91.12	90.37	3.20	3.45	5.05	5.22	0.63	0.96
	0.21	0.07	2.84	2.74	76.69	76.69	20.26	20.50	86.36	85.95	6.63	6.78	6.24	6.55	0.77	0.73
	0.12	0.12	6.08	6.06	73.56	73.58	20.24	20.24	77.07	76.87	13.27	13.23	8.72	8.84	0.94	1.06
	0.07	0.31	10.30	10.34	69.60	69.94	20.03	19.42	66.88	67.38	20.10	19.78	11.60	11.51	1.43	1.33
	0.64	1.15	14.94	15.17	65.56	65.22	18.86	18.46	57.37	57.67	25.80	25.38	14.89	15.00	1.94	1.96
	3.32	2.84	19.77	20.18	59.94	59.68	17.07	17.30	48.58	48.75	29.92	30.02	18.56	18.33	2.94	2.90

Potential Applications of Artificial Neural Networks to Thermodynamics

The deviations between experimental and predicated compositions in both phases are calculated according to equation (14). These values are compared with the calculated deviations from NRTL and UNIQUAC models (Goncalves et al., 2002). As be shown, GMDH model provided a better estimation against the other models.

$$\Delta w = 100 \sqrt{\frac{\sum_{n}^{N}\sum_{i}^{C}\left[\left(w_{i,n}^{I,ex} - w_{i,n}^{I,calc}\right)^2 + \left(w_{i,n}^{II,ex} - w_{i,n}^{II,calc}\right)^2\right]}{2NC}} \quad (14)$$

where N is the total number of tie lines, C is the total number of components. w is the mass fraction, the subscripts i, n are component and tie line, respectively and the superscripts I and II stand for oil and alcoholic phases, respectively; ex and calc refer to experimental and calculated concentrations.

In this work, we designed a GMDH model for different water concentration in solvent from all point of data set. For a better comparison with NRTL and UNIQUAC models (Goncalves et al., 2002) that were presented for systems with different water concentration, one can design four GMDH models for systems with 5, 8, 12, and 18% aqueous ethanol. It is obvious that each GMDH model uses four inputs $(X_2, ..., X_5)$ that are the mass fractions of the components according to Table 6.1.

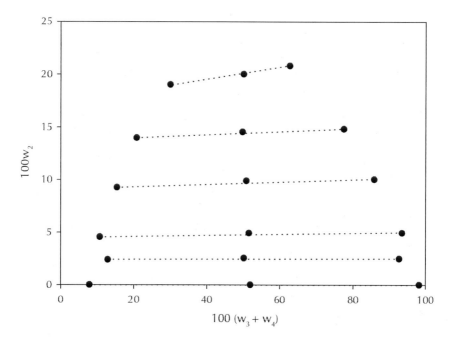

FIGURE 6.3 System of cyclohexane + ethanol + water at 298.15 K: (•) experimental; (...) GMDH.

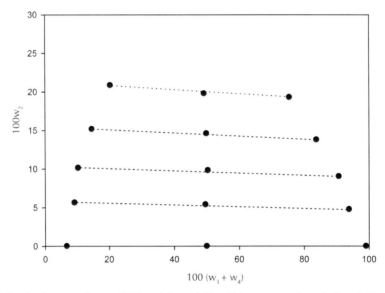

FIGURE 6.4 System of cron oil (1) + oleic acid (2) + 8% aqueous solvent [ethanol (3) + water (4)] at 298.15 K: (•) experimental; (...) GMDH.

6.3 CONCLUSION

In this study, a GMDH model designed using the experimental LLE data for system corn oil + oleic acid + ethanol + water at 298.15 K (Goncalves et al., 2002). The LLE data are predicted by GMDH model and then compared with the experimental data. Despite the complexity of the studied system, GMDH model allows a good prediction of phase equilibrium. Also the global deviation of the proposed model were lower than 0.57% in relation to the experimental data and the calculated data from NRTL an UNIQUAC models. GMDH model may be suitable to use in place of conventional methods predicting of LLE. The quality of the model is related to the quality of data used for the training of the model. For a better comparison it needs to design an independent GMDH model for each water concentration in solvent that can be studied in future works.

KEYWORDS

- Artificial neural network
- Group method of data handling
- Least-squares technique
- Liquid–liquid equilibrium
- Mass fractions
- Neural network structure
- Quaternary system

REFERENCES

Aksenova, T. I., Volkovich, V., and Villa, A. E. P. (2005). Robust Structural Modeling and Outlier Detection with GMDH-Type Polynomial Neural Networks, ICANN (2), 881–886.

Farlow, S. J. (1984). *Self-Organizing Method in Modeling: GMDH Type Algorithm.* Dekker, New York.

Ganguly, S. (2003)*Comput. Chem. Eng.* **27**, 1445–1454.

Goncalves, C. B., Batista, E., and Meirelles, A. J. A. (2002). Liquid–liquid equilibrium data for system cron oil + oleic acid + ethanol + water at 298.15 K. *J. Chem. Eng. Data* **47**, 416–420.

Guimaraes, P. R. B. and McGreavy, C. (1995). Flow of information through an artificial neural network, *Comput. Chem. Eng.* **19**(S1), 741–746.

Ivakhnenko, A. G. (1971). Polynomial theory of complex systems, *IEEE Trans. Syst. Man Cybern* **1**, 364–78.

Madala, H. R. and Ivakhnenko, A. G. (1994). *Inductive Learning Algorithms for Complex Systems Modeling.* CRC Press Inc., Boca Raton.

Mohanty, S. (2005). Estimation of vapor liquid equilibria of binary systems, carbon dioxide–ethyl caproate, ethyl caprylate and ethyl caprate using artificial neural networks, *Fluid Phase Equilib* **235**, 92–98.

Nariman-Zadeh, N., Darvizeh, A., Felezi, M. E., and Gharababaei, H. (2002). Polynomial modelling of explosive compaction process of metallic powders using GMDH-type neural networks and singular value decomposition, *Modelling Simul. Mater. Sci. Eng.* **10**, 727–744.

Onwubolu, G. C. (2007). Data Mining using Inductive Modelling Approach, International Workshop on Inductive Modelling-IWIM2007, 78–86.

Sharma, R., Singhal, D., Ghosh, R., and Dwivedi, A. (1999). Potential applications of artificial neural networks to thermodynamics: Vapor–liquid equilibrium predictions, *Comput. Chem. Eng.* **23**, 385–390.

Urata, S., Takada, A., Murata, J., Hiaki, T., and Sekiya, A. (2002). Prediction of vapor–liquid equilibrium for binary systems containing HFEs by using artificial neural network, *Fluid Phase Equilib* **199**, 63–78.

7 A Note on Application of Non-random Two-liquid (NRTL) Model

CONTENTS

7.1 Introduction 61
7.2 Experimental 62
Keywords 64
References 64

7.1 INTRODUCTION

Aqueous solutions containing salts are of increasing importance and influence on separation processes in chemical engineering. The electrolyte influence must be considered both in process design and operation, because it can significantly change the equilibrium composition. Aqueous liquid–liquid equilibrium is the results of intermolecular forces, mainly of the hydrogen-bonding type; addition of a salt to such systems introduces ionic forces that affect the thermodynamic equilibrium. When the mutual solubility decreases due the salt addition, the size of the two-phase region increases, and this effect is called "salting-out". In the opposite, when the solubility increases, the effect is called "salting-in". According to Al-Sahhaf and Kapetanovic (1997), the salting-in effect can be used to remove organic compounds from water.

Aqueous electrolyte liquid–liquid equilibrium is often related to extraction processes. For instance, the ethyl acetate recovery from its mixture with ethanol involves an aqueous extraction step in order to remove the ethanol. In this case, it is important to decrease the mutual solubilities of water and ester, improving the separation and yielding a dryer ester. In this particular case, Pai and Rao (1966) have studied the addition of sodium or potassium acetate for the ternary system water + ethanol + ethyl acetate. The alcohols used were 2-butanol and 3-methyl-1-butanol, and the salts used were sodium chloride, sodium acetate and calcium chloride. Santos et al. (1999) (Ghanadzadeh and Ghanadzadeh, 2002) studied the salt effect of potassium chloride and potassium sulfate on the water + ethanol + 1-pentanol system. In this work, experimental liquid–liquid equilibrium data for the ternary system water/2-butanol/acetone are determined at sodium acetate are added to the original ternary system, in order to determine the salting-out effect. Experimental tie-lines of the quaternary system water/2- water/2-butanol/acetone/ sodium acetate are also determined at the temperature.

Binary liquid–liquid equilibrium data for the system water + 2-butanol, ternary liquid–liquid equilibrium data for the system water + acetone + 2-butanol, and quaternary liquid–liquid equilibrium data for the system water + acetone + 1-butanol + sodium acetate were used for estimation of the energy interaction parameters of the NRTL model for the activity coefficient. The estimation procedures used the Aspen. With these parameters, the experimental data were correlated.

7.2 EXPERIMENTAL

All the reagents, 1-butanol, acetone, sodium acetate and sodium chloride were of analytical grade (Merck) and were used without further purification.

Experiments are carried out in equilibrium cells, such as those suggested by Ghanadzadeh (2002). The cell temperature is regulated by a thermostatic bath the overall mixture is prepared directly inside the cell, and the components are weighed on an analytical balance (Ohaus AS200, accurate to 0.0001 g). The mixture is vigorously agitated with a magnetic stirrer for 3 hr, in order to allow an intimate contact between the phases, and the equilibrium is achieved by letting the mixture rest for 12 hr. The system splits in two liquid phases that become clear and transparent at equilibrium, with a well defined interface. Separate samples of both phases are collected and analyzed.

Water, 2-butanol and acetone are determined by gas chromatography using a Varian CX 300 Star gas chromatograph with a Porapak-Q packed column and a thermal conductivity detector; the hydrogen flow rate was 30 cm^3 min^{-1} and the column temperature was 180°C. However, the salt cannot be allowed in the packed column or in the detector. So, an empty column section of 30 cm is placed before the packed column. The salt is deposed on the inner walls of this empty column, being eliminated from the gas stream. Periodically, this section is washed with distillated water and acetone, and further dried at 120°C.

In this way, the concentration of the volatile compounds is determined directly from chromatographic analysis. The salt concentration can be determined by gravimetrical analysis, evaporating the solution at 120°C, until constant mass. In this way, the mass of salt is determined. These two results can be combined in order to obtain the true mole fractions of the complete system; the compositions determined by chromatography are valid for the vaporized fraction, obtaining the mass of each component. By dividing the latter by the total mass, the true mass fraction (and hence the true mole fractions) are determined.

Vianna et al. (1992) validated this method when determining liquid–liquid equilibrium data of mixtures containing sodium acetate. The same Ohaus AS200 analytical balance was used to perform the gravimetrical analysis. All measurements are performed in triplicate.

A Note on Application of Non-random Two-liquid (NRTL) Model

TABLE 7.1 Liquid–liquid equilibrium of the water + 2-butanol + acetone + salt system at 25°C (mole fractions).

Aqueous phase (raffinate) mole fraction				Organic Phase (extract) mole fraction			
Water	2-butanol	Acetone	Sodium acetate	Water	2-butanol	Acetone	Sodium acetate
0.9750	0.0000	0.0250	0.000	0.5675	0.0195	0.4130	0.000
0.9688	0.0067	0.0218	0.000	0.5755	0.0253	0.3992	0.000
0.9672	0.0110	0.0218	0.000	0.5880	0.0360	0.3760	0.000
0.9610	0.0156	0.0234	0.000	0.5950	0.0453	0.3597	0.000
0.9591	0.0170	0.0239	0.000	0.6174	0.0581	0.3245	0.000
0.9501	0.0236	0.0263	0.000	0.6244	0.0643	0.3113	0.000
0.9431	0.0280	0.0289	0.000	0.6464	0.0693	0.2843	0.000
0.9366	0.0316	0.0318	0.000	0.6690	0.0724	0.2586	0.000
Sodium acetate (10 wt%)								
0.9423	0.0023	0.01	0.0454	0.6150	0.1425	0.2373	0.0052
0.9365	0.0086	0.0085	0.0464	0.4456	0.1404	0.4096	0.0044
0.9362	0.0119	0.0097	0.0422	0.4238	0.0828	0.4893	0.0041
0.9305	0.019	0.0094	0.0411	0.4220	0.0765	0.4975	0.0040
0.9239	0.0187	0.0107	0.0467	0.3966	0.0237	0.5759	0.0038
0.9423	0.0023	0.01	0.0454	0.6150	0.1425	0.2373	0.0052

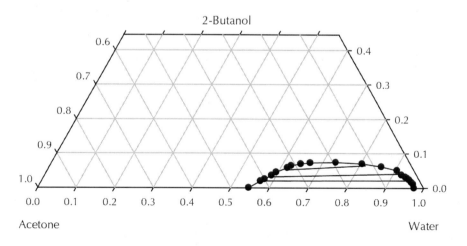

FIGURE 7.1 Experimental and calculated liquid–liquid equilibrium of the water + 1-butanol + acetone + salt system at 25°C.

KEYWORDS

- Cell temperature
- Chromatography
- Experimental
- Liquid–liquid equilibrium
- Non-random two-liquid

REFERENCES

Al-Sahhaf, T. A. and Kapetanovic, E. (1997). *J. Chem. Eng. Data* **42**, 74–77.

Ghanadzadeh, H. and Ghanadzadeh, A. (2002). *Fluid Phase Equilibria*, 202/2 339.

Pai, M. U. and Rao, K. M. (1966). *J. Chem. Eng. Data* **11**, 353–356.

Santos, G. R. d'Ávila, S. G., and Aznar, M. (20–24 June, 1999). V Conferencia Iberoamericana sobre Equil´ıbrio entre Fases para el Diseño de Processos, *EQUIFASE'99*, vol. 1, Vigo, Spain, pp. 193–201.

Vianna, R. F., d'Ávila, S. G., and Marinho, R. L. (9–11 September, 1992). In Proceedings of the IX Brazilian Congress of Chemical Engineering—COBEQ (in Portuguese), vol. 3, Salvador, Brazil, pp. 531–539.

8 Some Practical Hints on Application of UNIQUAC Solution Model

CONTENTS

8.1 Introduction 66
8.2 Experimental 66
    8.2.1 Chemicals 66
    8.2.2 Procedure 66
    8.2.3 The Uniquac Model 67
8.3 Discussion and Results 68
8.4 Conclusion 73
Keywords 73
References 74

NOMENCLATURES

a = Optimized interaction parameter
C = Number components
E = Excess property
q_i = Relative surface area per molecule
r_i = Number of segments per molecule
l_i = Pure–component constant

Root mean square deviation (rmsd%)
T = Absolute temperature (Kelvin)
u_{ij} = Interaction energy
Uniquac = Universal quasi chemical
x_i = Equilibrium mole fraction of component I
Z = Lattice coordination number, set equal to 10
z_i = Number of moles of component I

Greek Symbols
Φ = Segment fraction
θ = Area fraction
$τ_{ij}$ = Adjustable parameter in the UNIQUAC equation

Superscript
C = Combinatorial part of the activity coefficient
Q = UNIQUAC equation
R = Residual part of the activity coefficient

8.1 INTRODUCTION

Liquid–liquid equilibrium (LLE) data of multi component systems are important in both theoretical and industrial applications (design of many chemical processes and separation operations). The LLE investigations of ternary systems have been the subject of much interest in recent years (Alkandary et al., 2001; Al-Muhtaseb and Fahim, 1996; Arce et al., 1995; Fernandez-Torres et al., 1999; Garcia-Flores et al., 2001; Higashiuchi et al., 1995; Prausnitz et al., 1980; Wisniewska-Goclowska and Malanowski, 2001). Dilute ethanol solutions are not easily separated from water and, therefore, extraction of alcohols such as ethanol from aqueous mixtures is still an important problem. Various organic solvents for extraction of dilute ethanol solutions from aqueous mixtures have been investigated and reported in the literature (Dadgar and Foutch, 1985; Ghannadzadeh, 1993; Ghanadzadeh and Ghanadzadeh, 2002). However, from the experimental results, cyclohexane was chosen as the best solvent for recovering ethanol from aqueous mixtures (Ghannadzadeh, 1993).

A universal quasi–chemical model (UNIQUAC) (Abrams and Prausnitz, 1975) has been successfully applied for the correlation of several LLE systems. This model depends on optimized interaction parameters between each pair of components in the system, which can be obtained by experiments. The UNIQUAC equation can be fitted to the experimental composition by optimizing the interaction parameter (a_{ij} and a_{ji}). The optimized interaction parameters can also be correlated with temperature.

In this work, LLE data for (water + ethanol + cyclohexane) at temperatures ranging from 298.2 to 313.2 K were measured, and UNIQUAC model were used to correlate these data. The values for the interaction parameters were obtained for this equilibrium model. The effect of ethanol addition on solubility of water in cyclohexane (organic phase) was also investigated at different temperatures (298.2, 303.2, 308.2, and 313.2) K.

8.2 EXPERIMENTAL

8.2.1 Chemicals

Ethanol and cyclohexane were obtained from Merck at a purity of 99.8% and were used without further purification. The purity of these materials was checked by gas chromatography. Water was distilled before being used.

8.2.2 Procedure

A 300 ml glass cell connected to a thermostat was made to measure the LLE data. The temperature of the cell was controlled by a water jacket and maintained with an accuracy of within ± 0.1 K. A magnetic stirrer provided sufficient agitation within the apparatus. The prepared mixtures were introduced into the extraction cell and were stirred for 2 hr, and then left to settle for 4 hr for phase separation.

Samples of less than 1 µl were carefully taken by a syringe from the upper layer and trough a sampling tap from the lower layer. The Both phases were analyzed using Konik gas chromatography (GC) equipped with a thermal conductivity detector (TCD) and Shimadzu C-R2AX integrator. A 2 m × 2 mm column was used to separate the components. The injection temperature was 523.2 K and the detector temperature was 549.2 K. The carrier gas (helium) flow rate was maintained at 46 ml/min.

The internal standard method was used for the calibration of the TCD's response. In this study the internal standard was cyclohexane. Each of the measurements was carried out with several repetitions. The greatest error in mole fraction composition during calibration was about 0.0005.

8.2.3 The UNIQUAC Model

At LLE, the composition of the two phases (Raffinate phase and Extracted phase) can be determined from the following equations

$$(\gamma_i x_i)^1 = (\gamma_i x_i)^2 \tag{1}$$

$$\sum x_i^1 = \sum x_i^2 = 1 \tag{2}$$

Here γ_i^1 and γ_i^2 are the corresponding activity coefficients of component i in phase 1 and phase 2, x_i^1, and x_i^2 are the mole fractions of component i in the system and in phases 1 and 2, respectively. The interaction parameters between water, ethanol, cyclohexane are used to estimate the activity coefficients from the UNIQUAC. Equations 1 and 2 are solved for the mole fraction (x) of component i in the two liquid phases. This method of calculation gives a single tie line.

The UNIQUAC model (universal quasi-chemical model) is given by Abrams and Prausnitz (Abrams and Prausnitz, 1975).

$$\frac{g^E}{RT} = \sum_{i=1}^{c} x_i \ln(\frac{\Phi_i}{x_i}) + \frac{z}{2} \sum_{i=1}^{c} q_i x_i \ln(\frac{\theta_i}{\Phi_i}) - \sum_{i=1}^{c} q_i x_i \ln(\sum_{j=1}^{c} \theta_j \tau_{ji}) \tag{3}$$

or

$$\ln \gamma_i = \ln \gamma_i^c + \ln \gamma_i^R \tag{4}$$

where

$$\ln \gamma_i^c = \ln\left(\frac{\Phi_i}{x_i}\right) + \frac{z}{2} q_i \ln\left(\frac{\theta_i}{\Phi_i}\right) + l_i - \frac{\phi_i}{x_i} \sum_{j=1}^{c} x_j l_j \tag{5}$$

$$\ln \gamma_i^R = q_i \left[1 - \ln\left(\sum_{j=1}^{c} \theta_j \tau_{ji}\right) - \sum_{j=1}^{c} \frac{\theta_j \tau_{ij}}{\sum_{k=1}^{c} \theta_k \tau_{kj}} \right] \tag{6}$$

Here, γ_i^c is combinatorial part of the activity coefficient, and γ_i^R is the residual part of the activity coefficient. The τ_{ij} is adjustable parameter in the UNIQUAC equation. x_i is equilibrium mole fraction of component i. The parameter Φ_i (segment fraction) and θ_i (area fraction) are given by the following equations

$$\Phi_i = \frac{x_i r_i}{\sum_{i=1}^{c} x_i r_i} \tag{7}$$

$$\theta_i = \frac{x_i q_i}{\sum_{i=1}^{c} x_i q_i} \tag{8}$$

$$\tau_{ij} = \exp\left(-\frac{u_{ij} - u_{jj}}{RT}\right) \tag{9}$$

The parameter u_{ij} characterizes the interaction energy between compounds i and j and u_{ij} equals u_{ji}.

$$l_i = \left(\frac{z}{2}\right)(r_i - q_i) - (r_i - 1) \tag{10}$$

where, z is lattice coordination number, r_i is number of segments per molecule, q_i is relative surface area per molecule and l_i is pure—component constant (12).

8.3 DISCUSSION AND RESULTS

The experimental and calculated LLE data for the ternary system (water + ethanol + cyclohexane) at temperature of 298.15 K is graphically presented in Figure 7.1. The system exhibited type 1 phase behavior (Treybal, 1963), having only one liquid pair of partially miscible (cyclohexane + water) and two pairs of completely miscible (water + ethanol) and (ethanol+ cyclohexane).

The UNIQUAC model was used to correlate the experimental LLE data. As it can be seen from Figure 7.1, the predicted tie lines (dashed lines) are relatively in good agreement with the experimental data (solid lines). In other words, the UNIQUAC equations fairly fit the experimental data for this ternary system. The optimum UNIQUAC interaction parameters u_{ij} between water, ethanol, and cyclohexane were determined using the observed liquid–liquid data, where the interaction parameters describe the interaction energy between molecules i and j or between each pair of compounds.

Table 8.1 shows the calculated value of the UNIQUAC binary interaction parameters for the mixture water + ethanol + cyclohexane using universal values for the UNIQUAC structural parameters. The equilibrium model was optimized using an objective function, which was developed by Sørensen (1980).

Some Practical Hints on Application of UNIQUAC Solution Model

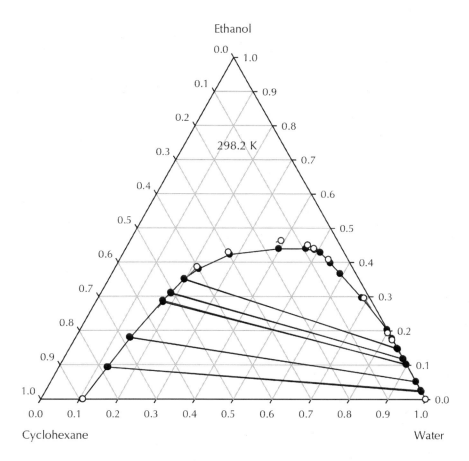

FIGURE 8.1 Experimental (—•) and predicted UNIQUAC (---o) LLE data at 298.2 K.

TABLE 8.1 The UNIQUAC binary interaction parameters (u12 and u21) optimized for (water + ethanol + cyclohexane), at 298.2 K.

Components	Water	Ethanol	cyclohexane
Water	0.00	-42.38	220.86
Ethanol	384.16	0.00	-46.41
Cyclohexane	228.75	152.13	0.00

The objective function obtained by minimizing the square of the difference between the mole fractions calculated by UNIQUAC model and the experimental data. The UNIQUAC structural parameters r and q were calculated from group contribution data that has been previously reported (Abrams and Prausnitz, 1975). The values of r and q used in the UNIQUAC equation are presented in Table 8.2.

TABLE 8.2 The UNIQUAC structural parameters.

Components	R	q
Ethanol	2.57	2.34
Cyclohexane	6.15	5.02
Water	0.92	1.40

The goodness of the fit, between the observed and calculated mole fractions, was calculated in terms of the root mean square deviation (rmsd). The rmsd values were calculated according to the equation of percentage root mean square deviations (rmsd%):

$$RMSD\% = 100 \sqrt{\sum_{k}^{n} \left[\frac{\sum_{i}^{3}\sum_{j}^{2}\left(x_{i,\exp}-x_{i,calc}\right)_{j}^{2}}{4n} \right]} \qquad (11)$$

where n is the number of tie-lines, $x_{\exp}$ indicates the experimental mole fraction, x_{calc} is the calculated mole fraction, and the subscript i indexes components, j phases and $k = 1,2,...n$ (tie lines). The average (rmsd%) between the observed and calculated mole fractions with a reasonable error was 1.70% (see Table 8.7). The experimental and predicted liquid–liquid equilibria data for the ternary system at different temperatures have been compiled in Tables 8.3–8.6.

The UNIQUAC model takes into account the effect of the temperature on the equilibrium with the term τ_{ij}:

$$\tau_{ij} = \exp(-a_{ij}/T) \qquad (12)$$

where a_{ij} is the UNIQUAC parameter with temperature independent and represents the energy interactions between an i-j pair of molecules.

TABLE 8.3 Experimental and predicted LLE for (water + ethanol + cyclohexane) at 298.2 K.

Aqueous phase				Organic Phase			
Mole fraction water		Mole fraction ethanol		Mole fraction water		Mole fraction ethanol	
Exp	Uniquac	Exp	Uniquac	Exp	Uniquac	Exp	Unique
0.9772	0.9799	0.0226	0.0198	0.2100	0.1812	0.1690	0.1668
0.9043	0.9062	0.0995	0.0963	0.2260	0.1966	0.2302	0.2019
0.8861	0.8869	0.1137	0.0816	0.2340	0.2385	0.2911	0.2858
0.8650	0.8548	0.1248	0.1229	0.2814	0.2728	0.3510	0.3202
0.8556	0.8749	0.1441	0.1241	0.3360	0.3120	0.3709	0.3644
0.8275	0.8305	0.1421	0.1433	0.4050	0.4211	0.4100	0.3818
0.7798	0.7720	0.1690	0.1659	0.5285	0.5299	0.3914	0.3822
rmsd%	0.89		1.45		1.94		1.97

Some Practical Hints on Application of UNIQUAC Solution Model

TABLE 8.4 Experimental and predicted LLE (water + ethanol + cyclohexane) at 303.2 K.

Aqueous phase				Organic Phase			
Mole fraction water		Mole fraction ethanol		Mole fraction water		Mole fraction ethanol	
Exp	Uniquac	Exp	Uniquac	Exp	Uniquac	Exp	Uniquac
0.9781	0.9790	0.0213	0.0203	0.2241	0.2240	0.1890	0.2241
0.9045	0.9059	0.0949	0.0931	0.2370	0.2889	0.3337	0.3699
0.8873	0.8862	0.1119	0.1126	0.2665	0.3070	0.3617	0.4020
0.8679	0.8726	0.1311	0.1259	0.2942	0.3196	0.3876	0.4191
0.8524	0.8549	0.1463	0.1432	0.3241	0.3360	0.4036	0.4361
0.8302	0.8304	0.1683	0.1669	0.3946	0.3584	0.4178	0.4518
0.7902	0.7934	0.2071	0.2025	0.4576	0.3911	0.4151	0.4631
rmsd%	0.25		0.30		3.94		3.72

TABLE 8.5 Experimental and predicted LLE for the ternary system water/ethanol/cyclohexane at 308.2 K.

Aqueous phase				Organic Phase			
Mole fraction water		Mole fraction ethanol		Mole fraction water		Mole fraction ethanol	
Exp	Uniquac	Exp	Uniquac	Exp	Uniquac	Exp	Uniquac
0.9782	0.9797	0.0211	0.0199	0.1919	0.2253	0.1606	0.1219
0.9176	0.9046	0.0818	0.0944	0.2302	0.2598	0.3442	0.3695
0.8958	0.8847	0.1034	0.1139	0.2616	0.2807	0.3815	0.4016
0.8791	0.8711	0.1198	0.1273	0.2851	0.3202	0.3958	0.4187
0.8598	0.8534	0.1390	0.1446	0.3213	0.3365	0.4297	0.4358
0.8295	0.8294	0.1690	0.1679	0.3949	0.3589	0.4377	0.4515
0.7891	0.7916	0.2082	0.2041	0.4580	0.4216	0.4451	0.4627
rmsd%	0.75		0.73		3.03		2.72

TABLE 8.6 Experimental and predicted LLE for the ternary system water/ethanol/cyclohexane at 313.2 K.

Aqueous phase				Organic Phase			
Mole fraction water		Mole fraction ethanol		Mole fraction water		Mole fraction ethanol	
Exp	Uniquac	Exp	Uniquac	Exp	Uniquac	Exp	Uniquac
0.9777	0.9785	0.0216	0.0250	0.1985	0.2001	0.1691	0.1683
0.9096	0.9046	0.0898	0.0943	0.2423	0.2890	0.3173	0.3097
0.8889	0.8848	0.1103	0.1138	0.2683	0.3069	0.3512	0.3920
0.8723	0.8712	0.1266	0.1271	0.2990	0.3194	0.3723	0.3891
0.8542	0.8537	0.1445	0.1443	0.3355	0.3357	0.3930	0.3998
0.8295	0.8201	0.1690	0.1598	0.3955	0.4010	0.4189	0.4191
0.7890	0.7980	0.2048	0.2101	0.4550	0.4512	0.4175	0.4362
rmsd%	0.55		0.47		2.43		1.85

TABLE 8.7 rmsd% values for the studied uniquac.

T/K	Uniquac
298.15	1.55
303.15	2.05
308.15	1.88
313.15	1.31
Average	1.70

The values of the interaction parameter (a_{ij} and a_{ji}) in the UNIQUAC equation with temperature dependent are shown in Table 8.8 (a_{ij} is expressed in Kelvin). It can be seen that, there is a linear increasing of the interaction parameters with increasing of temperature. Table 8.9 shows the values obtained for the interaction parameters assuming that there is a linear relation with temperature. These values were obtained from fitting to the values in Table 8.8.

TABLE 8.8 Optimized UNIQUAC binary aij for the system [water (1), ethanol (2), 2-etyl-1-hexanol (3)] at different temperatures (aij is expressed in Kelvin).

	298.15 K	303.15 K	308.15 K	313.15 K
a_{12}	205.53	202.40	197.43	193.76
a_{21}	-33.31	-29.40	-23.47	-18.83
a_{13}	208.9	207.26	205.94	204.12
a_{31}	215.11	221.30	227.10	233.52
a_{23}	-34.90	-38.70	-38.85	-42.12
a_{32}	106.84	111.3	114.80	118.60

TABLE 8.9 Optimized UNIQUAC binary interaction parameter a_{ij} (K) with dependence of temperature (T in Kelvin).

a_{ij} (K)	a_{ji} (K)
a_{12} = 446.01 - 0.8066T	a_{21} = -328.050 + 0.9874T
a_{13} = 302.28 - 0.3132T	a_{31} = -148.82 + 1.2206 T
a_{23} = 102.95 - 0.4622T	a_{32} = -124.180 + 0.7756T

The effect of ethanol addition on solubility of water in cyclohexane (organic phase) was also investigated at different temperatures 298.2, 303.2, 308.2, and 313.2 K Figure 7.2 shows that the solubility of water in cyclohexane increases with amounts of ethanol added to water + 2-ethyl-hexanol mixture. The temperature in the range of the study has only a small effect on the solubility of water in cyclohexane.

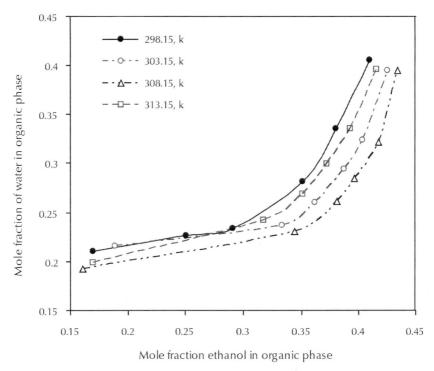

FIGURE 8.2 Effects of ethanol addition on solubility of water in cyclohexane at different temperatures.

8.4 CONCLUSION

The LLE data of the ternary system composed of water + ethanol + cyclohexane were measured at different temperatures of 298.2, 303.2, 308.2, and 313.2 K. The UNIQUAC model was used to correlate the experimental LLE data. The optimum UNIQUAC interaction parameters between water, ethanol, and 2-etyl-1-hexanol were determined using the experimental (liquid + liquid) data. The average rmsd value between the observed and calculated mole fractions with a reasonable error was 1.70% for the UNIQUAC model. The solubility of water in cyclohexane increases with amounts of ethanol added to water + 2-ethyl-hexanol mixture.

KEYWORDS

- **Cyclohexane**
- **Equilibrium model**
- **Gas chromatography**
- **Multi component systems**
- **Universal quasi–chemical model**

REFERENCES

Abrams, D. S. and Prausnitz, J. M. (1975). *AICHE* J. **21**, 116–128.

Al-Muhtaseb, S. A. and Fahim, M. A. (1996). *Fluid Phase Equilib.* 123, 189–203.

Alkandary, J. A., Aljimaz, A. S., Fandary, M. S., and Fahim, M. A. (2001). *Fluid Phase Equilib.* **187–188**, 131–138.

Arce, A., Blanco, A., Martinez-Ageitos, J., and Vidal, I. (1995). *Fluid Phase Equilib.* **109**, 291–297.

Dadgar, A. M. and Foutch, G. L. (1985). *Biotechnol. Prog.* **4**, 36–39.

Fernandez-Torres, M. J., Gomis-Yagues, V., Ramos-Nofuentes, M., and Ruiz-Bevia, F. (1999). *Fluid Phase Equilibr.* **164**, 267–273.

Garcia-Flores, B. E., Galicia-Aguilar, G., Eustaquio-Rincon, R., and Trejo, A. (2001). *Fluid Phase Equilib,* **185**, 275–293.

Ghannadzadeh, H. (1993). *Eleccion de disolventes selectivos para la extraccion en fase liquida de alcoholes $C_4(ABE)$ a partir de biomasa* Ph.D. Thesis, Universitat Politecnica de catalunya Barcelona, Spain.

Ghanadzadeh, H. and Ghanadzadeh, A. (2002). *Fluid Phase Equilib.* **202/2**, 337–344.

Higashiuchi, H., Sakuragi, Y., and Aria, Y. (1995). *Fluid Phase Equilib.* **110**, 197–204.

Prausnitz, J. M., Anderson, T. F., Grens, E. A., Eckert, C. A., Hsien, R., and Oconnell, J. P. (1980). *Computer Calculations for Multicomponent Vapor-Liquid and (liquid + liquid) Equilibria.* Prentice-Hall, Inc, Englewood.

Sorensen, J. M. (1980). *Correlation of liquid-liquid equilibrium data* Ph.D. Thesis, Technical University of Denmark, Lyngby, Denmark.

Treybal, R. E. (1963). *liquid-liquid Extraction*, 2nd Edition, McGraw-Hill Book Company, New York.

Wisniewska-Goclowska, B. and Malanowski, S. K. (2001). *Fluid Phase Equilibr.* **180**, 103–113.

9 Control of Liquid Membrane Separation Process

CONTENTS

- 9.1 Introduction 76
- 9.2 Bulk Liquid Membrane 76
- 9.3 Experimental 77
- 9.4 Parameters 78
- 9.5 Description of Model Experiment 78
  - 9.5.1 Feed Phase 78
  - 9.5.2 Stripping Phase 78
- 9.6 Error Analysis 78
- 9.7 Governing Equations In Transportation of Chromium 79
- 9.8 Diphenylcarbazide Method 80
- 9.9 Results 80
  - 9.9.1 Result of Feed Phase for Effect of pH 80
  - 9.9.2 Result of Stripping Phase for Effect of pH 80
  - 9.9.3 Result of Feed Phase for Effect of Percentage of Carrier in Membrane Phase 80
  - 9.9.4 Result of Stripping Phase for Effect of Percentage of Carrier in Membrane Phase 81
  - 9.9.5 Result of Feed Phase for Effect of Presence of Dodecanol in Membrane Phase 81
  - 9.9.6 Result of Stripping Phase for Effect of Presence of Dodecanol in Membrane Phase 81
  - 9.9.7 Result of Feed Phase for Effect of Velocity of Mixer in Stripping Phase 81
  - 9.9.8 Result of Stripping Phase for Effect of Velocity of Mixer in Stripping Phase 81
  - 9.9.9 Result of Feed Phase for Effect of Velocity of Mixer in Feed Phase 82
  - 9.9.10 Result of Stripping Phase for Effect of Velocity of Mixer in Feed Phase 82
  - 9.9.11 Result of Feed Phase for Effect of Molarity of Stripping Phase 82
  - 9.9.12 Result of Stripping Phase For Effect of Molarity of Stripping Phase 82
- 9.10 Discussion 83
  - 9.10.1 Effect of Presence/Absence of Surfactant in Membrane Phase 83
  - 9.10.2 Effect of pH 85
  - 9.10.3 Effect of Volumetric Percentage of Carrier 85
  - 9.10.4 Effect of Mixing 85

    9.10.5   Effect of Molarity of Stripping Phase 87
9.11 Conclusion 88
Keywords 88
References 89

NOMENCLATURES

if = Liquid environment near the feed phase
of = Organic environment near the feed phase
os = Organic environment near the stripping phase
is = Liquid environment near the stripping phase
o = Organic phase
i = Liquid phase
δ = Thickness of mass transfer film
D = Mass transfer diffusion coefficient
N = Mass transfer diffusion

9.1 INTRODUCTION

Membrane separation is an area deserving special attention because of its great potential for low capital cost and energy efficiency. To date, however, few membrane processes other than reverse osmosis and hydrogen separation have demonstrated any industrial utility primarily because of problems of speed and selectivity in separation (Douglas Way et al., 1982).

    Liquid membranes with impressive properties such as high selectivity and efficient consumption of energy in separation processes seem to be more suitable. Other advantages such as variety of configuration and carriers for different applications, simplicity of assembling and high rate of mass transfer are facilitated the implementation of these membranes. Chromium is one of contaminants that exist in waste water of various industries like steel, pigment, and leather tanning. In this project separation of Cu (II) ion by implementation of a bulk liquid membrane using alamine as carrier, kerosene as solvent, sodium hydroxide as stripping product phase, dodecanol for preventing from jellying of inorganic and organic phases have been investigated. Effective parameters on separation of Cu (II) ion including feed phase pH, stripping phase molarity, mixer rotational rate in feed and stripping phase, volume percentage of carrier in organic phase, presence/absence of surfactant in organic phase have been studied. In the range of designed experimental, the optimum conditions as follow have been found: pH = 2.45, stripping phase molarity = 3, mixer rotational rate in both inorganic phases = 100 rpm, volumetric amount of carrier = 1%, and presence of surfactant.

9.2 BULK LIQUID MEMBRANE

This set up is useful only for laboratory experiments, and is setup as follows. Following Figure 9.1, a U-tube cell is used, and some type of carrier, perhaps dissolved in CH_2Cl_2, is placed in the bottom of the tube. That is the organic membrane phase.

Two aqueous phases are placed in the arms of the U-tube, floating on top of the organic membrane. With a magnetic stirrer rotating at fairly slow speeds, in the range of 100–300 rpm, the transported amounts of materials are determined by the concentrations in the receiving phase. Stability is maintained so long as the stirrer is not spinning too quickly.

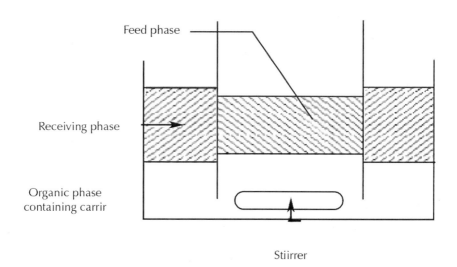

9.3 EXPERIMENTAL

For doing experiments we used glass cell which divided to two compartments by thin sheet of glass. Dimensions of cell were 10cm × 24cm × 20cm. Volumes of one section were three times greater than other section. In bottom of each section and 2 cm upper then the bottom two holes were arranged for sampling. Both holes sealed with plastic caps. Feed phase and stripping phase were pour in greater and little section respectively. Membrane phase was pour over foregoing both phases. Two pirex blades rotated by mixer with constant speed 15cm × 1cm diameter. Each one of blades have two paddles. Potassium chromate and NaOH were pouring with determined concentration in grate and little section of cell respectively. Membrane phase was pour over two foregoing phases in two steps. First, Kerosene was added without carrier (for example 350 ml), then 41 ml kerosene + 5 ml alcohol + 4 cc alamine was added to membrane phase. To avoid, of vaporization of membrane phase top of the cell was covered with foil. Sampling was started step by step up to 24 hr. (1 day).At the beginning of experiments, the color of feed phase was yellow. Membrane phase and stripping phase were color less. Gradually the color of feed phase approached to pale yellow and simultaneously the color of membrane and stripping phases approached to very pale yellow (colorless) and the color of stripping phase approached to dark yellow. This variation in the color of feed and stripping phases showed the process of extraction of chromium ion from feed phase to stripping phase. During the experiments, the pH of feed phase was controlled and sulfuric acid was added to feed phase for maintaining the pH.

9.4 PARAMETERS

Parameters were investigated as follow:

(1) pH of feed phase.
(2) Percentage of carrier in membrane phase.
(3) Presence/absence of dodecanol in membrane phase
(4) Molarity of stripping phase
(5) Velocity of mixer in feed phase
(6) Velocity of mixer in stripping phase

Where pH was investigated in three states like, 2/4/6, Presence of dodecanol in membrane phase was investigated in three states like, 0.5, 1, and 2%, Molarity of stripping phase was investigated in three states like: 1/2/3, velocity of mixer in both liquid phases was investigated in three states like: 80, 100, and 120 rpm.

9.5 DESCRIPTION OF MODEL EXPERIMENT

9.5.1 Feed Phase

Solving 1.94 g potassium chromate (K_2CrO_4) in 2 l DM water, (0.005 mol). Adding 5 ml sulfuric acid for approaching to solution with pH = 2. Diluting 100 ml of foregoing solution to 1 l with DM water, 750 ml of final solution was our feed phase.

9.5.2 Stripping Phase

Solving 30 g NaOH on 250 ml DM water.

9.6 ERROR ANALYSIS

For calculation of ions concentration error equation (1) was used:

$$\bar{X} \pm \frac{ts}{\sqrt{N}} \qquad (1)$$

where:

$\bar{X}$ = Average of 3 amounts of ion concentration obtained from calibration curve.
T = 95% safety limit with constant amount = 4.3
N = obtained samples
S = Standard deviation where

$$S = \sqrt{\frac{\sum(\bar{x} - x_i)^2}{N-1}}$$

in standard deviation equation we have $\bar{X}$ and x_i:

$\bar{X}$ and x_i are average of three amounts of ion concentration and ion concentration in each sampling respectively.

9.7 GOVERNING EQUATIONS IN TRANSPORTATION OF CHROMIUM

(a) Diffusion of chromium from bulk of feed to interface of feed phase and membrane phase:

$$N_{CrO_4^{2-}} = \frac{D_f}{\delta_{if}}\left([CrO_4^{2-}]_f - [CrO_4^{2-}]_{if}\right)$$

(b) Reaction on the interface of membrane phase and feed phase:

$$CrO_4^{-2}(if) + 2Al_2OH_{(of)} \leftrightarrow Al_2CrO_4^{(of)} + 2OH^{-(if)} \qquad (2)$$

$$K = \frac{[Al_2CrO_4]_{if}[OH^-]_{if}^2}{[AlOH]_{of}[CrO_4^{-2}]_{if}} \qquad (3)$$

and following reaction:

$$H^+ + OH^- \leftrightarrow H_2O \qquad K = \frac{1}{K_w} = 10^{14} \qquad (4)$$

(c) Diffusion of product complex of up reaction from interface between feed phase and membrane phase to bulk of membrane:

$$N_{Al_2CrO_4} = \frac{D_o}{\delta_{of}}\left([Al_2CrO_4]_{of} - [Al_2CrO_4]_o\right) \qquad (5)$$

(d) Diffusion of product complex of reaction (2) from bulk of membrane to interface of membrane and stripping phase:

$$N_{Al_2}CrO_4 = \frac{Do}{Dos}\left([Al_2CrO_4]_o - [Al_2CrO_4]os\right) \qquad (6)$$

(e) Reaction in the interface of membrane and stripping phases:

$$Al_2CrO_4(os) + 2OH^-(is) \leftrightarrow 2AlOH(os) + CrO_4^{-2}(is) \qquad (7)$$

$$K = \frac{[AlOH]_{os}[CrO_4^{-2}]_{is}}{[Al_2CrO_4]_{os}[OH^-]_{is}^2} \qquad (8)$$

and following reaction:

$$H^+ + OH^- \leftrightarrow H_2O \qquad K = \frac{1}{k_w} = 10^{14} \qquad (9)$$

(f) Diffusion of chromate ion to bulk of stripping phase:

$$NCrO_4^{-2} = \frac{Ds}{Dis}\left([CrO_4^{-2}]_{is} - [CrO_4^{-2}]_s\right) \qquad (10)$$

9.8 DIPHENYLCARBAZIDE METHOD

Diphenylcarbazide (*sym-* diphenylcarbazide, diphenylcarbohydrazide) reacts in acid medium with Cu (II) ions to give a violet solution which is the basis of this sensitive method.

Many investigators have studied the reaction (Bose, 1954), offering rather divergent explanations of its mechanism. Pflaum and Howick (1956), among others (Kovalenko and Petrashen, 1963; Minczewski and Zemijewska, 1960; Zittel, 1963), have shown that the cationic (II) and diphenylcarbazone fails to yield a violet color. In all probability, the reaction involves unhydrated chromium (III) ions formed during the oxidation of diphenylcarbazide to diphenylcarbazone.

This explanation is, however, incomplete since, when the colored reaction product is extracted into isoamyl alcohol or chloroform in the presence of perchlorate, the remaining colorless aqueous phase contains half of the chromium (Lichtenstein and Allen, 1959, 1961; Sano, 1962). When studying the reactions of diphenylcarbazide and diphenylcarbazone with various metal cations, Balt and van Dalen (1961) found that diphenylcarbazide only forms metal chelates after its oxidation to diphenylcarbazone.

9.9 RESULTS

9.9.1 Result of Feed Phase for Effect of pH

(Molarity of stripping phase = 3, rotational speed of mixer in all phases = 100 rpm, volumetric percentage of carrier = 1%, with presence of dodecanol in membrane phase)

Time (min)	Concentration (ppm) pH=2	Concentration (ppm) pH=4	Concentration (ppm) pH=6
1440	0.1 ± 0.0	0.3 ± 0.2	0.5 ± 0.2

9.9.2 Result of Stripping Phase for Effect of pH

(Molarity of stripping phase = 3, rotational speed of mixer in all phases = 100 rpm, volumetric percentage of carrier = 1%, with presence of dodecanol in membrane phase)

Time (min)	Concentration (ppm) pH=2	Concentration (ppm) pH=4	Concentration (ppm) pH=6
1440	70.4 ± 0.2	66.4 ± 0.8	58.6± 0.6

9.9.3 Result of Feed Phase for Effect of Percentage of Carrier in Membrane Phase

(Molarity of stripping phase = 3M, rotational speed of mixer in all phases = 100 rpm, pH = 2, with presence of dodecanol in membrane phase)

Time (min)	Concentration (ppm) % carrier (v/v)=0.5	Concentration (ppm) % carrier (v/v)=1	Concentration (ppm) % carrier (v/v)=2
1440	5.6 ± 0.5	0.1 ± 0.0	0

Control of Liquid Membrane Separation Process

9.9.4 Result of Stripping Phase for Effect of Percentage of Carrier in Membrane Phase

(Molarity of stripping phase = 3M, rotational speed of mixer in all phases = 100 rpm, pH = 2, with presence of dodecanol in membrane phase)

Time (min)	Concentration (ppm) % carrier (v/v)=0.5	Concentration (ppm) % carrier (v/v)=1	Concentration (ppm) % carrier (v/v)=2
1440	60.3 ± 0.4	70.4 ± 0.2	53.3 ± 0.5

9.9.5 Result of Feed Phase for Effect of Presence of Dodecanol in Membrane Phase

(Molarity of stripping phase = 3M, rotational Speed of mixer in all phases = 100 rpm, pH = 2, % Carrier (v/v) = 1)

Time (min)	Concentration (ppm) with presence of dodecanol in membrane	Concentration with out presence of dodecanol in membrane phase
1440	0.1 ± 0.0	6.4 ± 0.4

9.9.6 Result of Stripping Phase for Effect of Presence of Dodecanol in Membrane Phase

(Molarity of stripping phase = 3M, rotational speed of mixer in all phases = 100 rpm, pH = 2, % carrier (v/v)=1)

Time (min)	Concentration (ppm) with presence of dodecanol in membrane	Concentration(ppm) without presence of dodecanol in membrane phase
1440	70.4 ± 0.2	52.6 ± 0.9

9.9.7 Result of Feed Phase For Effect of Velocity of Mixer in Stripping Phase

(Molarity of stripping phase=3, rotational speed of mixer in feed phase = 100 rpm, volumetric percentage of carrier = 1% with presence of dodecanol in membrane phase, pH = 2)

Time (min)	Concentration(ppm) velocity of mixer in stripping phase =80rpm	Concentration(ppm) velocity of mixer in stripping phase =100rpm	Concentration(ppm) velocity of mixer in stripping phase =120rpm
1440	5.7 ± 0.5	0.1. ± 0.0	3.7 ± 0.4

9.9.8 Result of Stripping Phase For Effect of Velocity of Mixer in Stripping Phase

(Molarity of stripping phase = 3, rotational speed of mixer in feed phase = 100 rpm, volumetric percentage of carrier = 1% with presence of dodecanol in membrane phase, pH = 2)

Time (min)	Concentration(ppm) velocity of mixer in stripping phase =80rpm	Concentration(ppm) velocity of mixer in stripping phase =100rpm	Concentration(ppm) velocity of mixer in stripping phase =120rpm
1440	56.4 ± 0.9	70.4. ± 0.2	60.6 ± 0.9

9.9.9 Result of Feed Phase for Effect of Velocity of Mixer in Feed Phase

(Molarity of stripping phase = 3, rotational speed of mixer in stripping phase = 100 rpm, pH = 2, volumetric percentage of carrier=1% with presence of dodecanol in membrane phase.)

Time (min)	Concentration(ppm) velocity of mixer in stripping phase =80rpm	Concentration(ppm) velocity of mixer in stripping phase =100rpm	Concentration(ppm) velocity of mixer in stripping phase =120rpm
1440	6.2 ± 0.4	0.1. ± 0.0	3.0 ± 0.4

9.9.10 Result of Stripping Phase for Effect of Velocity of Mixer in Feed Phase

(Molarity of stripping phase = 3 rotational speed of mixer in stripping phase = 100 rpm, pH = 2, volumetric percentage of carrier = 1%, with presence of dodecanol in membrane phase.)

Time (min)	Concentration(ppm) velocity of mixer in feed phase =80rpm	Concentration(ppm) velocity of mixer in feed phase =100rpm	Concentration(ppm) velocity of mixer in feed phase =120rpm
1440	54.7 ± 0.5	70.4. ± 0.2	57.1 ± 0.7

9.9.11 Result of Feed Phase for Effect of Molarity of Stripping Phase

(rotational speed of mixer in all phases = 100 rpm, pH = 2/volumetric percentage of carrier = 1% with presence of do decanol in membrane phase.)

Time (min)	Concentration(ppm) 1M Solution	Concentration(ppm) 2M Solution	Concentration(ppm) 3M Solution
1440	0.2 ± 0.7	6.2. ± 0.4	6.2 ± 0.4

9.9.12 Result of Stripping Phase For Effect of Molarity of Stripping Phase

(rotational speed of mixer in all phases = 100 rpm, pH = 2/volumetric percentage of carrier = 1% with presence of dodecanol in membrane phase.)

Time (min)	Concentration(ppm) 1M Solution	Concentration(ppm) 2M Solution	Concentration(ppm) 3M Solution
1440	56.5 ± 0.7	66.5. ± 0.6	70.4 ± 0.2

Chromium accumulation % in Membrane phase (C.A) $= \dfrac{A-(B-C)}{A}$

A = initial moles of chromium in feed phase
B = final moles of chromium in feed phase
C = final moles of chromium in stripping phase

Parameter	Amount	C.A
	2	10.66
pH of feed phase	4	14.44
	6	22.44
	0.5	4.44
% Carrier(v/v)	1.0	10.66
	2.0	33.33

Presence/absence of dodecanol in membrane phase	YES	10.66
	NO	8.88
Molarity of stripping phase	1	26.66
	2	14.00
	3	10.66
Velocity of mixer in feed phase	80	4.44
	100	10.66
	120	15.55
Velocity of mixer in stripping phase	80	15.55
	100	10.66
	120	20.88

Chromium extraction percentage for each parameter

Extraction% = C/A *100

where

A = Initial moles of chromium in feed phase

C = Final moles of chromium in stripping phase

Parameter	Amount	Extraction %
pH of feed phase	2	88.88
	4	74.44
	6	71.11
Velocity of mixer in feed phase	80	71.11
	100	88.88
	120	73.33
Velocity of mixer in stripping phase	80	64.44
	100	88.88
	120	77.77
Molarity of stripping phase	1	71.11
	2	84.44
	3	88.88
Presence/absence of dodecanol in membrane phase	YES	88.88
	NO	66.66
% Carrier(v/v)	0.5	75.55
	1.0	88.88
	2.0	66.66

9.10 DISCUSSION

9.10.1 Effect of Presence/Absence of Surfactant in Membrane Phase

Presence of surfactant in membrane phase decreases surface tension between liquor and organic phases and because of this reason, mass transfer increases. Surfactant should be ineffective, has a high solubility in organic phase and has a polarity in one end. The 5 ml dodecanol was used in experiments. Using high amounts of this material create jelling state in surface of liquor and organic phases. Also when high amounts of dodecanol were used, Alamine salts: surrounded by dodecanol and cannot react immediately, therefore extraction decreases. As we see in Figure 9.1; After 24 hr in our experiment with presence of dodecanol, concentration of chromium in feed phase is 0.1 ppm and concentration of chromium in the experiment without presence of dodecanol after 24 hr is 6.4 ppm. Also in Figure 9.2: Concentration of chromium in stripping phase with presence of dodecanol is 70.4 ppm.

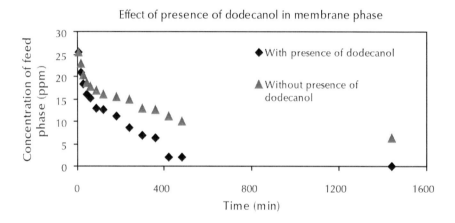

FIGURE 9.1

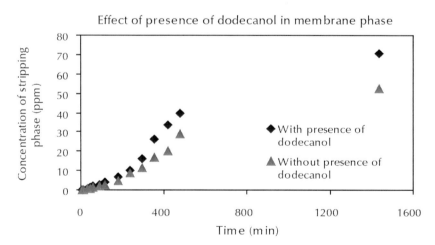

FIGURE 9.2

9.10.2 Effect of pH

Decreasing of pH in feed phase cause to neutralize OH⁻ in equation (2) and reaction goes to right side: it increases speed of this formula. Therefore if pH was bounded on 2, membrane efficiency increases. If pH was decreased so much may be membrane and carrier was oxidized, because of oxidizing property of chromate.

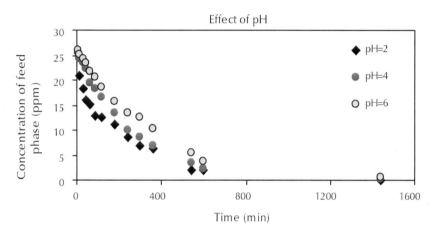

FIGURE 9.3

9.10.3 Effect of Volumetric Percentage of Carrier

With increasing of % carrier in membrane phase, extraction increases and in definite concentration of carrier (1%) maximum percentage of extraction was obtained. But after that, because of increasing in amount, in membrane phase and also increasing of carrier, extraction decreases. Then optimum amount of carrier should be defined. With increasing of % carrier in membrane phase, extraction was operated successfully but because of increasing of chromium in membrane phase, back extraction (extraction from organic phase to stripping phase) was operated slowly. As we considered % carrier was played important role in extraction process. Reaction (2); is faster than reaction (7); therefore amount of entering chromium to membrane phase is more than amount of balcony chromium from membrane phase. Since chromium decreases in feed phase in one an exact time speed of entering and balconing chromium equal to each other and amount of chromium in membrane phase goes maximum. Then optimum carrier percentage is so important in extraction processes, and in our experiments it was 1% (V/V).

9.10.4 Effect of Mixing

The speed of mixing of phases is so important that the fastest speed of mixing of phases lead to smallest thickness of films in equations (1), (5), (6), and (10) and because of this reason, mass transfer decreases. As we see in diagrams, the optimum speed of mixing is 100 rpm. Because in some other speed like 80 rpm, flow is laminar and it is so silent and phases cannot mix together as they can, and mass transfer do not operate successfully. Because of these reason extraction decreases.

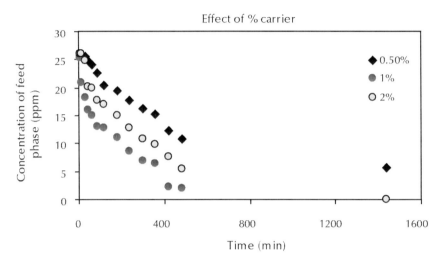

FIGURE 9.4

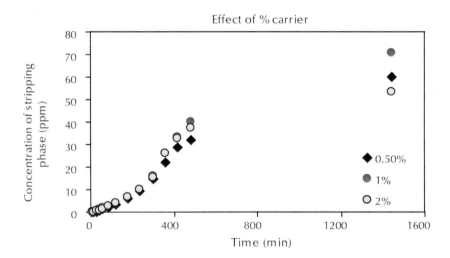

FIGURE 9.5

The next speed of mixing of phases that we investigated was 120 rpm. In 120 rpm, flow is turbulent and because of this reason, fine bubbles of feed and stripping phases enter to membrane phase and because of high velocity of mixers these bubbles cannot operate inter facial reactions and go straight to the other liquor phase and extraction decreases. As we see in Figure 9.6. concentration of feed phase after 24 hr, in 100, 80, and 120 rpm approaches to 0.1, 6.2, and 3 ppm respectively. Also, in Figure 9.7, extraction in 100 rpm, is higher than the others. And after 24 hr concentration of stripping phase in 100, 80, and 120 rpm is 70.4, 56.4, and 60.6 ppm respectively.

Control of Liquid Membrane Separation Process

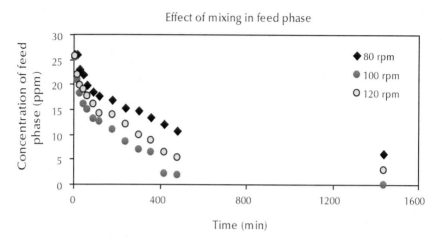

FIGURE 9.6

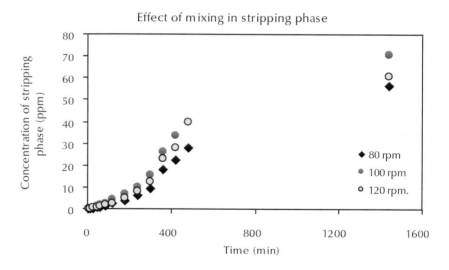

FIGURE 9.7

9.10.5 Effect of Molarity of Stripping Phase

Because of low concentration of chromium in feed phase and pay attention to this point that with each chromate ion, two H^+ ions were transferred from feed phase to stripping phase variation of pH in stripping phase is negligible. Therefore, most basic stripping phase lead to operate extraction easily. And we see this point in the diagram and results.

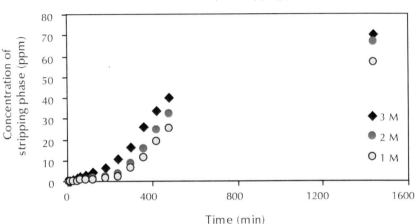

FIGURE 9.8

9.11 CONCLUSION

- Presence of surfactant in membrane phase increased extraction process.
- Optimum amount of carrier percentage in our experiments was 1% (V/V). In this measure, maximum extraction was obtained.
- Optimum amount of pH for feed in our experiments was pH = 2. In this pH, extractions are increased and Al_2CrO_4 complexes are generated faster than the other conditions that we did.
- Optimum velocity of mixer for both liquor phases (feed phase and stripping phase) in our experiments is 100 rpm. Because extraction and back extraction are operated successfully.
- Best molarity for stripping phase in our experiments is molarity = 3 because under this condition back extraction are operated successfully and AlOH complexes are generated faster than the other conditions that we did.

KEYWORDS

- **Diphenylcarbazide**
- **Feed Phase**
- **Membrane phase**
- **Membrane separation**
- **Stripping phase**

REFERENCES

Allen, T. L. (1958). *Anal. Chem.* **30**, 447.

Babko, A. K. and Get'man, T. E. (1959). *Zh. Obshch. Khim.* **29**, 2416.

Balt, S. and Van Dalen, E. (1961). *Anal. Chim. Acta* **25**, 507.

Bose, M. (1954). Ibid. 10,201,209.

Douglas Way, J., Richard D. Noble, Thomas M. Flynn, and Dendy Sloan, E. (1982). *Journal of Membrane Science* **12**, 239–259.

Goddard, J. D. (1977). Further applications of carrier mediated transport theory—A survey. *Chem. Eng. Sci.* **32,** 795.

Halwachs, W and Schugerl, R. (1980). The liquid membrane technique—A promising extraction process. *Int. Chem. Eng.* **20,** 519.

Kemula, W., Kublik, Z., and Najdeker, E. (1962). *Roczniki Chem.* **36**, 937.

Kimura, S. G., Matson, S. L., and Ward, W. J. (1979). III Industrial Applications of Facilitated Transport. In *Recent Developments in Separation Science*, Vol. 5. N. N. Li (Ed.). CRC Press, Cleveland, Ohio.

Kovalenko, E. V., and Petrashen, V. I. (1963). *Zh. Analit. Khim.* **18**, 743.

Lichtenstein, I. E. and Allen, T. L. *ibid* 81, 1040 (1959). *ibid* (1961). *J. Phys. Chem.***65**, 1238.

Minczewski, J. and Zemijewska, W. (1960). *Roczniki Chem.* **34**, 1559 *ibid* (1960). *Chem. Anal.* (Warsaw) 5,429.

Pflaum, R. T and Howick, L. C. (1956). *J. Am. Chem. Soc.* **78**, 4862.

Sano, H. (1962). *Anal. Chim. Acta* **27**, 398.

Schultz, J. S., Goddard, J. D., and Suchdeo, S. R. (1974a). Facilitated transport via carrier-mediated diffusion in membranes, Part I. *AIChE J.* **20**, 417.

Schultz, J. S., Goddard, J. D., and Suchdeo, S. R. (1974b). Facilitated transport via carrier-mediated diffusion in membranes, Part II. *AIChE J.* **20**, 625.

Smith, D. R., Lander R. J., and Quinn, J. A. (1977). *Carrier-Mediated Transport in Synthetic Membranes*. In *Recent Developments in Separation Science*, Vol. 3, N. N. Li (Ed.). CRC Press, Cleveland, Ohio.

Zittel, H. E. (1963). *Anal. Chem.* **35**, 329.

10 Development of Artificial Neural Network (ANN) Model for Estimation of Vapor Liquid Equilibrium (VLE) Data

CONTENTS

10.1 Introduction 91
10.2 Artificial Neural Network Theory 92
    10.2.1 The Multi-Layer Perceptron (Mlp) Network 93
10.3 Neural Network Model 94
10.4 Discussion And Results 95
10.5 Conclusion 98
Keywords 99
References 99

10.1 INTRODUCTION

The precise vapor–liquid equilibrium (VLE) data of binary mixtures like alcohol–alcohol are important to design many chemical processes and separation operations. The VLE investigations of binary systems have been the subject of much interest in recent years (Artigas et al., 1997; Artigas et al., 2001; Hiaki et al., 2002; Lliuta et al., 2000; Monton et al., 2005; Oracz et al., 1996; Rodriguez et al., 2002; Seo et al., 2000; Vecher et al., 2005).

Conventional method of estimating the VLE is based on equations of state (EOS). These EOS although derived from strong theoretical principles and involve a number of adjustable parameters in terms of binary interaction parameters, as well as parameters in mixing rule equations. Furthermore, the binary interaction parameters that are functions of both temperature and composition need to be calculated at every temperature at which the VLE is required. The iterative method of estimation of VLE using EOS makes it unsuitable for real time control. The development of numerical tools, such as neural networks, has paved the way for alternative methods to estimate the VLE (Ganguly, 2003; Guimaraes and McGreavy, 1995; Petersen et al., 1994; Sharma et al., 1999; Urata et al., 2002). It has attracted considerable interest because of its ability to capture with relative ease the non-linear relationship between the independent and dependent variables. Several authors have reported application of artificial neural network (ANN) for estimation of thermodynamic properties such as estimation of viscosity, density, vapor pressure, compressibility factor, and VLE. An ANN model for estimation of vapor pressure from aerosol composition, relative humidity, and temperature has been reported by Potuchuti and Wexler. Chouai et al. (1997) have used an ANN model for estimating the compressibility factor for the liquid and vapor phase as a function of temperature and pressure for several refrigerants. The ANN has also been used for estimating the shape factors as a function of temperature and density for a number of refrigerants that can be used in the extended corresponding state model (Chouai et al., 2002; Scalabrin et al., 2002). Lagier and Richon (2002) have used ANN model for estimation of compressibility factor and density as a function of pressure and temperature for some refrigerants. Although, a number of papers have been published with experimental data for VLE for various systems and estimation of VLE using conventional thermodynamic models, not many have used this technique for estimating the VLE. An ANN based group contribution method for estimation of liquid phase activity coefficient have been suggested by Petersen et al. (1994) that can be used for estimation of VLE. A multilayer perceptron with a single hidden layer has been used by Guimaraes and McGreavy (1995) for estimating the VLE of benzene–hexane system. Sharma et al. (1999) have used the multi-layer perceptron (MLP) model to estimate the VLE for the methane–ethane and ammonia–water systems. They have also highlighted the advantage of ANN over conventional EOS for estimating the VLE systems containing polar compounds. Ganguly (2003) on the other hand, has used the radial basis function to estimate the VLE for several binary and ternary systems. Urata et al. (2002) have estimated the VLE using two multi-layer perceptrons. The input parameters for the first ANN are normal boiling point divided by molecular weight, density, and dipole moment for both the components and the output is a negative or positive sign. The second ANN has an extra input of mole fraction of one of the components in the liquid phase in addition to the inputs of the first ANN. The output from the second ANN is logarithm of the activity coefficient for that component. Using the logarithmic activity coefficients, vapor liquid composition, and equilibrium temperature were estimated. Mohanty (2006) has used a single multilayer perceptron for estimating the VLE of carbon dioxide–difluoromethane system.

In this chapter, attempt has been made to use ANN for estimating the VLE for the systems tert-butanol+1-hexanoland n-butanol+2-ethyl-1-hexanol. In the next section, the theory of ANN has been explained and the type of ANN which is used for estimating the VLE has been defined. In section 3, the data inputs to the network has been shown at atmospheric pressure and in temperature range of 353.2-458.2 K within the results and in section 4, the outputs of ANN has been compared with experimental results.

10.2 ARTIFICIAL NEURAL NETWORK THEORY

The driving force behind the development of the ANN models is the biological neural network, a complex structure, which is the information processing system for a living being. Thus ANN mimics a human brain for solving complex problems, which may be otherwise difficult to solve using available mathematical techniques. The advantage of using an ANN model is that it does not require any other data except the input and output data. Once the model has been adequately trained, the input data is sufficient to estimate the output. The other advantage is a single model can be used to get multiple outputs. From its initiation in the early forties till today there are hundreds of ANN architecture developed, however, there are a few such as MLP and radial basis function that are more popular and find wide applications. Details have been dealt with elsewhere (Bishop, 1994; Bishop, 1995), therefore only a brief description of multilayer perceptron neural network that belongs to the feed forward neural network architecture in general has been described.

10.2.1 The Multi-Layer Perceptron (MLP) Network

This type of network is composed of an input layer, an output layer, and one or more hidden layers (Figure 10.1). Bias term in each layer is analogous to the constant term of any polynomial. The number of neurons in the input and the output layer depends on the respective number of input and output parameters taken into consideration. However, the hidden layer may contain zero or more neurons. All the layers are interconnected as shown in the figure and the strength of these interconnections is determined by the weights associated with them. The output from a neuron in the hidden layer is the transformation of the weighted sum of output from the input layers and is given as (1)

$$z_j = g\left(\sum_{i=0}^{d} w_{ji} p_i\right) \quad (1)$$

The output from the neuron in the output layer is the transformation of the weighted sum of output from the hidden layer and is given as (2)

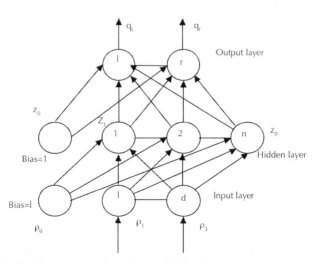

FIGURE 10.1 Multilayer perception with one hidden layer.

$$qk = \tilde{g}\left(\sum_{j=0}^{n} \tilde{w}_{ji} z_i\right) \qquad (2)$$

where p_i is the *i*th output from the input layer, z_j is the *j*th output from the hidden layer w_{ij} is the weight in the first layer connecting neuron *i* in the input layer to neuron *j* in the hidden layer, $\tilde{w}\ kj$ is the weight in the second layer connecting neuron *j* in the hidden layer to the neuron *k* in the output layer and *g* and $\tilde{g}$ are the transformation functions. The transformation function is usually a sigmoid function with the most common being (3),

$$g(a) = \tanh a = \frac{e^a - e^{-a}}{e^a + e^{-a}} \qquad (3)$$

The other commonly used function is (4),

One of the reasons for using these transformation functions is the ease of evaluating the derivatives that is required for minimization of the error function.

10.3 NEURAL NETWORK MODEL

The neural network model for the two binary systems viz. *tert*-butanol+1-hexanoland n-butanol+1-hexanolis based on the experimental data reported by Ghanadzadeh et al. (2005). The summary of the data is shown in Table 10.1 and 10.2. All neural networks take numeric input and produce numeric output. The transformation function of a neuron is typically chosen so that it can accept input in any range, and produce output in a strictly limited range. Although the input can be in any range, there is a saturation effect so that the unit is only sensitive to inputs within a fairly limited range. Numeric values have to be scaled into a range that is appropriate for the network.

The three input parameters to the multi-layer perceptron are the atmospheric pressure and the mole fraction of liquid (X1), and vapor (Y1) phases. The output parameter is the boiling temperature.

TABLE 10.1 The experimental data of VLE for ethanol + water.

T (K)	X_1	Y_1	T (K)	X_1	Y_1
373	0.0000	0.0000	352	0.5714	0.6879
361	0.0714	0.3912	352	0.6429	0.7225
358	0.1429	0.4976	352	0.7143	0.7624
356	0.2143	0.5476	351	0.7857	0.8088
355	0.2857	0.5796	351	0.8571	0.8628
354	0.3571	0.6057	351	0.9286	0.9259
353	0.4286	0.6309	351	1.0000	1.0000

TABLE 10.2 The experimental data of VLE for acetone + water.

T (K)	X_1	Y_1	T (K)	X_1	Y_1
373	0.0000	0.0000	332	0.5000	0.8491
345	0.0714	0.6888	332	0.5714	0.8533
337	0.1429	0.7865	331	0.6429	0.8596
334	0.2143	0.8194	331	0.7143	0.8694
333	0.2857	0.8339	330	0.7857	0.8846
332	0.3571	0.8412	330	0.8571	0.9080
332	0.4286	0.8455	330	0.9286	0.9439

At first, this network has been learned by the experimental inputs and output. During the training period, optimizing the weights minimizes the error between the experimental and estimated boiling temperature. The derivatives of the error function with respect to the weights are estimated using the error back propagation technique, in which the error in the output layer is propagated backwards to estimate the derivatives in the lower layer (Bishop, 1994). The minimization of the error function is then carried out using the gradient descent method in which the weights are moved in the direction of negative gradient. Varying the number of neurons in the hidden layer carries out training. The model with the minimum number of neurons in the hidden layer that gives the desired accuracy is selected. A single hidden layer was found to be sufficient for all the three cases.

10.4 DISCUSSION AND RESULTS

Two neural networks have been used in this research. In the first network, the ANN input data is the mole fractions of liquid and vapor phases and the output is the activity coefficient of binary system. The experimental data and the estimated results of the activity coefficient are given in Table 10.3 and 10.4.

TABLE 10.3 Experimental and estimated of the activity coefficient for ethanol + water.

Exp	ANN	Exp	ANN
5.6970	6.0000	1	0.9170
3.7538	4.3622	1.0141	0.9849
2.7521	3.2517	1.0514	1.0578
2.1596	2.5394	1.1075	1.1361
1.7838	2.1138	1.1797	1.2203
1.0122	1.3672	1.2664	1.3106
1.0028	1.0021	2.3900	2.3208
1.0000	1.0011	2.5750	2.4927

TABLE 10.4 Experimental and estimated of the activity coefficient for acetone + water.

EXP	ANN	EXP	ANN
8.6156	8.4700	1.0000	1.0000
5.8192	6.0209	0.8900	1.0121
4.2179	4.3162	1.0477	0.8900
3.2290	3.1793	1.1066	0.9200
2.5757	2.4567	1.1904	1.0400
2.1230	2.0186	1.3023	1.0700
1.7988	1.7587	1.4480	1.2900

In the Figures 10.4 and 10.5, the experimental data and the ANN output have been compared. Then, the activity coefficients, which are output of the first network together with the atmospheric pressure, mole fraction of liquid, and vapor phases, are given to the second network. After the learning and training of ANN the output, which is temperature, generated. Now, we can compare the experimental data with the output of ANN. Figures 10.2 and 10.3 show this comparison. The output data of model is given in the Table 10.3 and 10.4. The average absolute deviation for the temperature output was in range of 2-3. % and for the activity coefficient was less than 0.001%. If more experimental data are available for the present system, the model could be improved to be applicable for a much wider range.

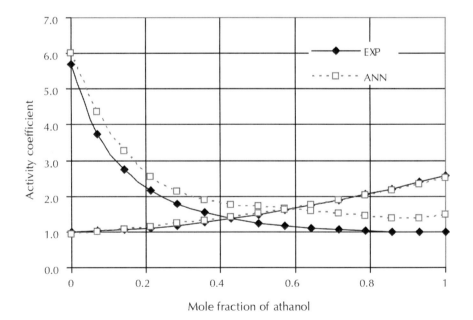

FIGURE 10.2 The activity coefficient of ethanol + water.

Development of Artificial Neural Network (ANN) Model for Estimation

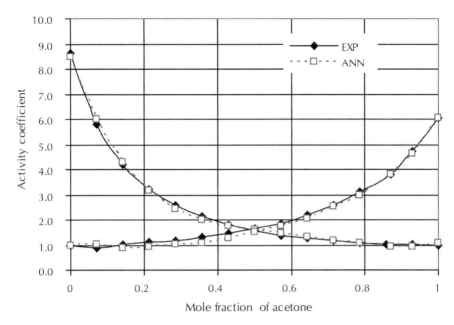

FIGURE 10.3 The activity coefficient of acetone + water.

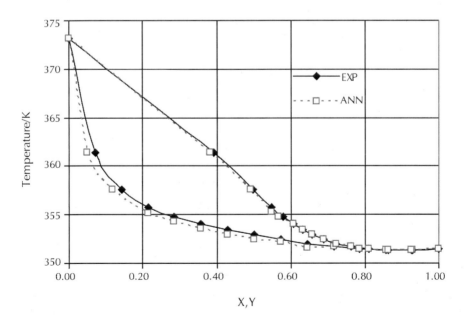

FIGURE 10.4 Boiling temperature diagram (T) for the system of ethanol + water.

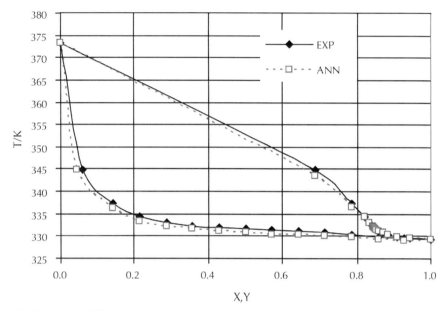

FIGURE 10.5 Boiling temperature diagram (T) for the system acetone + water.

10.5 CONCLUSION

Development of ANN model for estimating VLE is less cumbersome than methods based on EOS. It does not require parameters such as the critical properties of the components or the binary interaction parameters, nor the mixing rules as required by conventional methods. Binary interaction parameters may not be linearly related to the temperature and hence assumption of linear relation may lead to erroneous results. Once the ANN model is trained estimation of VLE is a one step process. This considerably saves computational time. Hence, it may be highly suitable to use it in place of conventional methods for real time process control. Since ANN works like a black box, it can be applied to any type of binary mixture for which the VLE data is available irrespective of the type of the system. However, the major disadvantage of this technique is that it can be used only in the range in which it has been trained, as it is empirical in nature. In this work, artificial neural network models have been developed for the binary systems, acetone+water, and ethanol+water VLE in the temperature range of is 330.2–375.2K and the atmospheric pressure. The weights have been optimized so as to minimize the error between the estimated and experimental VLE. The weights for the models have been tabulated for all the binary systems that can be used for predicting the VLE at any temperature. The models were able to estimate the VLE satisfactorily. The percent deviation in estimating the vapor phase mole fraction was found to be similar to experimental data in ANN model. The average absolute deviation for the boiling temperature was in range of 2–3% and for the activity coefficient was less than 0.001.The weights thus optimized during the training period can be used in ANN models for predicting the VLE of the binary systems at the boiling temperature in the range considered in this chapter.

KEYWORDS

- Carbon dioxide–difluoromethane system
- Equations of state
- Multi-layer perceptron
- Vapor–liquid equilibrium

REFERENCES

Artigas, H. Lafuente, C., Lopez, M. C., Royo, F. M., and Urieta, J. S. (1997). *Fluid Phase Equilib.* **134**, 163.

Artigas, H., Lafuente, C., Martin, S., Minones, J. Jr., and Royo, F. M. (2001)., *Fluid Phase Equilib.* **192**, 49.

Bishop, C. M. (1994). *Rev. Sci. Instrum.* **65**, 1803–1832.

Bishop, C. M. (1995). *Neural Networks for Pattern Recognition.* Oxford University Press, Oxford.

Chouai, A., Laugier, S., and Richon, D. (2002). *Fluid Phase Equilib.* **199**, 53–62.

Ganguly, S. (2003). *Comput. Chem. Eng.* **27**, 1445–1454.

Ghanadzadeh, H., Ghanadzadeh, A., Sariri, R., and Boshra, A. (2005). *Fluid Phase Equilib.* **233**, 123-128.

Guimaraes, P. R. B. and McGreavy, C. (1995). *Comput. Chem. Eng.* **19**(S1), 741–746.

Hiaki, T., Nanao, M., Urata, S., and Murata, J. (2002). *Fluid Phase Equilib.* **194**, 969.

Laugier, S. and Richon, D. (2003). *Fluid Phase Equilib.* **210**, 247–255.

Lliuta, M. C., Lliuta, I., and Lavachi, F. (2000). *Chem. Eng. Sci.* **55**, 2813.

Mohanty, S. (2006). *Int. J. Refrigeration* **29**, 243.

Monton, J. B., Munyoz, R., Burguet, M. C., and de la Torre, J. (2005). *Fluid Phase Equilib.* **227**, 19.

Oracz, P., Goral, M., Wilczek Vera, G., and Warycha, S. (1996). *Fluid Phase Equilib.* **126**, 71.

Petersen, R., Fredenslund, A., and Rasmussen, P. (1994). *Comput. Chem. Eng.* **18**, s63–67.

Potukuchi, W. and Wexler, A. S. (1997). *Atmospheric Environ.* **31**, 741–753.

Rodriguez, A., Canosa, J., Domenguez, A., and Tojo, J. (2002). *Fluid Phase Equilib.* **198**, 95.

Scalabrin, G. Piazza, L., and Richon, D. (2002). *Fluid Phase Equilib.* **199**, 33–51.

Scalabrin, G., Piazza, L., and Cristofoli, G. (2002). *Int. J. hermophys.* **23**, 57–75.

Seo, J. Canosa, J., Lee, J., and Kim, H. (2000). *Fluid Phase Equilib.* **172**, 211.

Sharma, R., Singhal, D., Ghosh, R., and Dwivedi, A. (1999). *Comput. Chem. Eng.* **23**, 385–390.

Urata, S., Takada, A., Murata, J., Hiaki, T., and Sekiya, A. (2002). *Fluid Phase Equilib.* **199**, 63–78.

Vecher, E., Vicent, A., Gonzales, R., and Marteniz, A. (2005). *Fluid Phase Equilib.* **227**, 239.

11 Some Aspects of a Fluid Phase Equilibria and UNIFAC Model

CONTENTS

11.1 Introduction 102
11.2 Experimental 102
    11.2.1 The Unifac Model 103
11.3 Results And Discussion 105
11.4 Conclusion 112
Keywords 112
References 112

NOMENCLATURE

a = Optimized interaction parameter
C = Number components
E = Excess property is pure–component constant
n = Number of tie-line
q_i = Relative surface area per molecule
r_i = Number of segments per molecule
RMSD = Root mean square deviation %
T = Absolute temperature (Kelvin)
u_{ij} = Interaction energy
Unifac = Universal quasi chemical
x_i = Equilibrium mole fraction of component i
Z = Lattice coordination number, set equal to 10
z_i = Number of moles of component i

Greek Symbols

F = Segment fraction
q = Area fraction
τ_{ij} = Adjustable parameter in the UNIFAC equation
l_i = Pure–component constant.

Superscript
C = Combinatorial part of the activity coefficient
Q = UNIFAC equation
R = Residual part of the activity coefficient

11.1 INTRODUCTION

Liquid–liquid equilibrium (LLE) data of multi-component systems are important in both theoretical and industrial applications design of many chemical processes and separation operations. The LLE investigations of ternary systems have been the subject of much interest in recent years (Aljimaz et al., 2000; Arce et al., 1995; Arce et al., 2001; Garcia-Floreset al., 2001; Garcia et al., 1988; Ghanadzadeh and Ghanadzadeh, 2003; .Ince and Ismail Kirbaslar, 2003; Jassal et al., 1994). Dilute formic acid solutions are not easily separated from water and, therefore, extraction of formic acid from aqueous mixtures is still an important problem. Various organic solvents for extraction of dilute formic acid solutions from aqueous mixtures have been investigated and reported in the literature (Kollerup and Daugulis, 1985; Pesche and Sandler, 1995; Zhang and Hill, 1991). However, from the experimental results, cumene was chosen as the best solvent for recovering formic acid from aqueous mixtures (Ghanadzadeh and Ghanadzadeh, 2003).

The UNIFAC model defines functional groups, which make up the structures of compounds. Each functional group makes a unique contribution to the compound property. The interaction parameters obtained for a small number of groups using thermodynamically consistent data can be used for multi-component systems.

The UNIFAC (the universal quasi-chemical function group activity coefficient) is one of the best methods in estimating activity coefficient that has been established to date (Fredenslund et al., 1975; Fredenslund et al., 1977; Magnussen et al., 1981) has been successfully applied for the prediction of several LLE systems. This model depends on interaction parameters between each pair of components in the system, which can be obtained by between each of the main groups.

In this work, LLE data for water + formic acid + cumene at temperatures ranging from 298.2 to 313.2 K were measured, and UNIFAC model were used to estimate these data. The values for the interaction parameters were obtained for this equilibrium model. The effect of formic acid addition on solubility of water in cumene (organic phase) was also investigated at different temperatures 298.2, 303.2, 308.2, and 313.2 K.

11.2 EXPERIMENTAL

The schematic diagram of the LLE apparatus is similar to that of Peschke and Sandler (1995). The equilibrium cell is made of 250 ml glass cell connected to a thermostat was made to measure the LLE data. The temperature of the cell was controlled by a water jacket and maintained with an accuracy of within ± 0.1 K. The temperature was measured using a calibrated digital thermometer traceable to NIST. The liquid mixture in the cell was agitated vigorously by a magnetic stirrer. For each run, the agitation was continued at least 4 hr to sufficiently mix the compounds. The mixture was then settled at least 8 hr to completely separate the two liquid phases. The sample of organic-rich phase was taken with a syringe (1 μl) from the top opening of the cell and that of water-rich phase from a sampling port at the bottom of the cell. This method can avoid cross contamination by the other phase during the sampling procedure. The composition of sample was analyzed using Konik gas chromatography (GC) equipped with a thermal conductivity detector (TCD) and Shimadzu C-R2AX integrator. A stainless steel column packed with 10% Porapak QS 80/100 (2m x 1/8 in.). The injection temperature was 524.2 K and the detector temperature was 551.2 K. The carrier gas (helium) flow rate was maintained at 50 ml/min can clearly separate the constituent compounds of the samples. Four samples were replicated for each phase at a fixed experimental condition. In general, the repeatability of the area fractions is about ± 0.02%. The averaged area fraction was converted into mole fraction by the calibration equations. Calibrations were made with gravimetrically prepared samples in two composition range, in accordance with those in the water-rich and organic-rich phases for each aqueous binary system. Refractive indexes were measured with CETI refract meter, its stated accuracy is ±0.05%. Densities were measured using a DA-210 (Kyoto electronic) digital density meter, in combination with a remote measuring cell and a calibrated thermometer. The instrument was initially calibrated using distilled HPLC grade water ($\rho = 997.15 \pm 0.00001$ kg m^{-3} at 293.15 K) and air. According to the manufacturer's procedure, the estimated uncertainty in the density measurement was ± 0.01% kg m^{-3}. The temperature of the density meter was controlled by circulation of water and measured with a copper-constantan thermocouple. The measured physical properties together with literature data[1] (Weast, 1989) are presented in Table 11.1.

Formic acid and cumene were obtained from Merck at a purity of 99.8% and were used without further purification. The purity of these materials was checked by gas chromatography. Deionized distilled water was prepared in our laboratory.

TABLE 11.1 Physical properties of the components at T=298.2 K and P=101.325 kpa.

Property	Water Exp	Water Lit	Formic acid Exp	Formic acid Lit	Cumene Exp	Cumene Lit
ρ(kg.m^{-3})	999.85	997.04	1218.50	1220	863.3	864.
n_{od}	1.3326	1.3325	1.3712	1.3714	1.4913	1.4915
$T_{b(K)}$	373.20	373.30	383.2	383.15	425.400	425.56

11.2.1 The UNIFAC Model

The UNIFAC method, where in activity coeficients in mixtures are related to interactions between structural groups. (1820). At LLE, the composition of the two phases (Raffinate phase and Extracted phase) can be determined from the following equations:

$$(\gamma_i x_i)^1 = (\gamma_i x_i)^2 \tag{1}$$

$$\sum x_i^1 = \sum x_i^2 = 1 \tag{2}$$

Here, γ_i^1 and γ_i^2 are the corresponding activity coefficients of component i in phase 1 and phase 2, x_i^1, and x_i^2 are the mole fractions of component i in the system and in phases 1 and 2, respectively. The LLE data of the ternary mixture were predicted by UNIFAC method (Fredenslund et al., 1975; Fredenslund et al., 1977; Magnussen et al., 1981) using the interaction parameters between CH_3, CH_2,CH, OH, and H_2O obtained by Magnussen et al. (1981) and Fredenslund et al. (1975). The values of the UNIFAC parameters for LLE predictions are summarized in Table 11.2.

Formic acid, cumene are used to estimate the activity coefficients from the UNIFAC. Equations 1 and 2 are solved for the mole fraction (x) of component i in the two liquid phases. This method of calculation gives a single tie line.

The UNIFAC model is given by Magnussen and et al. (Fredenslund et al., 1975; Fredenslund et al., 1977; Magnussen et al., 1981)

$$\ln \gamma_i^c = \ln\left(\frac{\Phi_i}{x_i}\right) + \frac{z}{2} q_i \ln\left(\frac{\theta_i}{\Phi_i}\right) + l_i - \frac{\phi_i}{x_i} \sum_{j=1}^{c} x_j l_j$$

$$l_i = \left(\frac{z}{2}\right)(r_i - q_i) - (r_i - 1)$$

where, z is lattice coordination number, r_i is number of segments per molecule; q_i is relative surface area per molecule and l_i is pure component constant (Fredenslund et al., 1975; Fredenslund et al., 1977; Magnussen et al., 1981). Here, γ_i^c is combinatorial part of the activity coefficient. The parameter Φ_i (segment fraction) and θ_i (area fraction) are given by the following equations:

$$\Phi_i = \frac{x_i r_i}{\sum_{i=1}^{c} x_i r_i} \qquad (3)$$

$$\theta_i = \frac{x_i q_i}{\sum_{i=1}^{c} x_i q_i} \qquad (4)$$

$$\ln \Gamma_k = Q_k \left[1 - \ln\left(\sum_{m=1}^{c} \theta_m \tau_{mk}\right) - \sum_{m=1}^{c} \left(\frac{\theta_m \tau_{km}}{\sum_{n=1}^{c} \theta_n \tau_{nm}}\right) \right] \qquad (5)$$

$$\ln \gamma_i^R = \sum_k v_k^i \left(\ln \Gamma_k - \ln \Gamma_k^i\right) \qquad (6)$$

γ_i^R is the residual part of the activity coefficient. τ_{ln} is adjustable parameter in the UNIFAC equation. X_i is equilibrium mole fraction of component I. Where E_{ke} is the group residual activity coefficient and Γ_k^i is the residual activity coefficient of group k in a reference solution containing only molecules of type i. (Fredenslund et al., 1977).

or

$$\ln \gamma_i = \ln \gamma_i^c + \ln \gamma_i^R \qquad (7)$$

where

$$\theta_m = \frac{x_m Q_m}{\sum_{n=1}^c x_n Q_n} \qquad (8)$$

$$X_m = \frac{v_m^i x_i}{\sum_i \sum_n v_n^i x_i} \qquad (9)$$

where n, m, and k are counters for the UNIFAC groups and the counters i and j are used for different compounds in the system. θ_m and x_m are group surface area and group fraction. The group interaction parameter τ_{mk} is given by

$$\hat{O}_{mk} = \exp\left(-\frac{a_{mk}}{T}\right) \qquad (10)$$

$$r_i = \sum_k i_k^i R_k \qquad (11)$$

$$q_i = \sum_k i_k^i Q_k \qquad (12)$$

where v_k^i, always an integer, is the number of groups of type k in molecule i. Group parameters R_k and Q_k are obtained from the van der Waals group volume and surface areas (Fredenslund et al., 1977). Parameters r_i and q_i are calculated as the sum of the group volume and area.

TABLE 11.2 UNIFAC group parameters for prediction tie-lines data (Fredenslund et al., 1977).

Name	-CH$_3$	-CH$_2$	-ACH	-ACCH	COOH	H$_2$O
-CH$_3$	0.	0.	61.13	737.5	663.5	1318
-CH$_2$	0.	0.	61.13	737.5	663.5	1318
-ACH	-11.12	-11.12	0.	167	537.4	903.8
-ACCH	-69.70	-69.70	-146.8	0.	603.8	5696.
-COOH	315.4	315.4	62.32	-138.4	0.	-292.0
H$_2$O	580.6	580.6	362.8	377.6	225.4	0.

11.3 RESULTS AND DISCUSSION

The experimental and calculated LLE data for the ternary system water + formic acid + cumene at temperatures of 300.2315.2 K presented in Table 11.3 and are graphically presented in Figures. 11.2, 11.3, 11.4 and 11.5. The system exhibited type 1 phase behavior (Pesche and Sandler, 1995), having only one liquid pair of partially miscible (cumene + water) and two pairs of completely miscible (water + formic acid) and (formic acid + trans decalin). The solubility of water in cumene at different temperatures is presented in Figure 11.1.

The solubility of water in cumene, 2EH, heptanol, hexanol, and dimethyl glutarate in Figure 11.1 shows that solubility of water in cumene is smaller than 2EH, heptanol, and hexanol for that reason this solvent is suitable for process liquid–liquid extraction.

The UNIFAC model and experimental LLE data: As it can be seen from Figures 11.2, 11.3, 11.4, and 11.5, the predicted tie lines (dashed lines) are relatively in good agreement with the experimental data (solid lines). In other words, water + formic acid + cumene using universal values for the UNIFAC structural parameters calculated from group contribution data that has been previously reported (Aljimaz et al., 2000; Ghanadzadeh and Ghanadzadeh, 2003; Ince and Ismail Kirbaslar, 2003).

This show the selectivity and strength of the solvent in extracting the acid, distribution coefficients (D_i) for the formic acid (i = 2) and water (i = 1), and the separation factor (S) are determined as follows:

$$D_i = \frac{\text{mole fraction in solvent phase}\,(x_{i3})}{\text{mole fraction in aqueous phase}\,(x_{i1})} \tag{13}$$

$$S = \frac{D_2}{D_1} \tag{14}$$

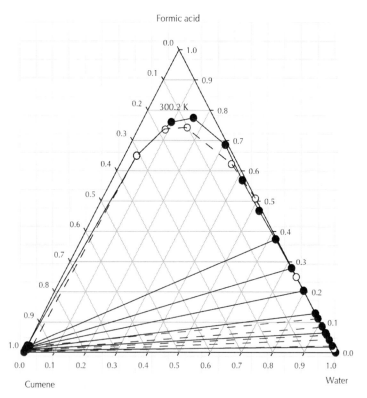

FIGURE 11.1 Experimental (—●—) and predicted UNIFAC (---○---) LLE data at 300.2 K.

Some Aspects of a Fluid Phase Equilibria and UNIFAC Model

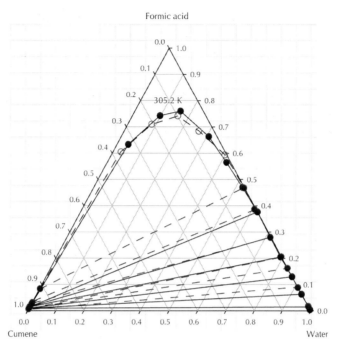

FIGURE 11.2 Experimental (—●) and predicted UNIFAC (---○) LLE data at 305.2 K.

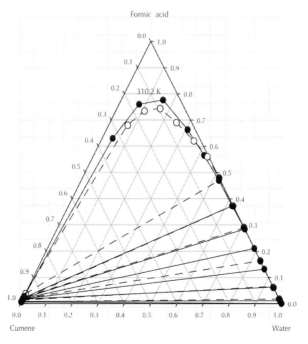

FIGURE 11.3 Experimental (—●) and predicted UNIFAC (---○) LLE data at 310.2 K.

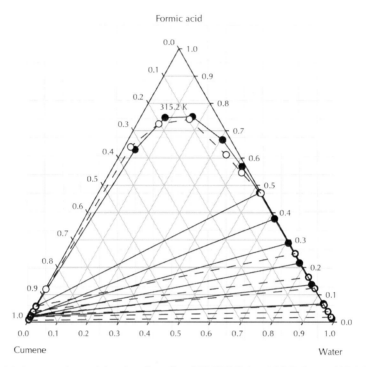

FIGURE 11.4 Experimental (—•) and predicted UNIFAC (---o) LLE data at 315.2 K

TABLE 11.3 Experimental tie lines of the system, together with the rmsd% values in the prediction by UNIFAC for the ternary system water (1) + formic acid (2) + cumene (3).

| Aqueous phase (raffinate) mole fraction |||| Organic Phase (extract) mole fraction ||||
| X_1 (water) || X_2 (formic acid) || X_1 (water) || X_2 (formic acid) ||
Exp.	UNIFAC	Exp.	UNIFAC	Exp.	UNIFAC	Exp.	UNIFAC
\multicolumn{8}{c}{300.2 K}							
0.9822	0.9889	0.0177	0.0111	0.0001	0.0004	0.0052	0.0066
0.9358	0.9785	0.0642	0.0215	0.0002	0.0008	0.0088	0.0071
0.8712	0.9594	0.1287	0.0406	0.0003	0.0009	0.0099	0.0090
0.7962	0.9426	0.2035	0.0574	0.0003	0.0009	0.0131	0.0120
0.7207	0.9145	0.2780	0.0854	0.0003	0.0085	0.0142	0.0089
0.6226	0.8879	0.3736	0.1120	0.0003	0.0091	0.0174	0.0097
0.5238	0.7502	0.4669	0.2488	0.0003	0.0098	0.0212	0.0201
0.4190	0.4910	0.5684	0.5080	0.0004	0.0399	0.6489	0.6489
0.3069	0.3580	0.6850	0.6212	0.0399	0.0890	0.0052	0.7350
rmsd%=8.75								

TABLE 11.3 (Continued)

Aqueous phase (raffinate) mole fraction				Organic Phase (extract) mole fraction			
X_1 (water)		X_2 (formic acid)		X_1 (water)		X_2 (formic acid)	
Exp.	UNIFAC	Exp.	UNIFAC	Exp.	UNIFAC	Exp.	UNIFAC
305.2 K							
0.9956	0.9827	0.0071	0.0173	0.0001	0.0001	0.0043	0.0078
0.9854	0.9112	0.0084	0.0888	0.0002	0.0002	0.0145	0.0097
0.9365	0.8375	0.0087	0.1625	0.0003	0.0005	0.0634	0.0096
0.8699	0.7939	0.0095	0.2060	0.0003	0.0008	0.1299	0.0111
0.7912	0.7178	0.0143	0.2791	0.0003	0.0009	0.2053	0.0271
0.7168	0.6070	0.0152	0.3870	0.0003	0.0011	0.2801	0.0317
0.6231	0.5300	0.0133	0.4680	0.0003	0.0030	0.3769	0.1999
0.5243	0.4810	0.0124	0.5120	0.0004	0.0290	0.4698	0.6050
0.4195	0.3650	0.6337	0.6300	0.0404	0.0842	0.5651	0.7100
rmsd%=8.59								
310.2 K							
0.9993	0.9825	0.0055	0.0076	0.0000	0.0001	0.0007	0.0076
0.9885	0.9388	0.0077	0.0080	0.0000	0.0002	0.0115	0.0080
0.9368	0.8367	0.0163	0.0173	0.0001	0.0004	0.0631	0.0173
0.8681	0.7095	0.0167	0.0179	0.0001	0.0005	0.1318	0.0179
0.7892	0.6271	0.0180	0.0171	0.0001	0.0008	0.2107	0.0171
0.7155	0.5205	0.0195	0.0290	0.0001	0.0009	0.2844	0.0290
0.6232	0.4350	0.0204	0.0376	0.0037	0.0012	0.3731	0.0376
0.5244	0.3530	0.0284	0.6800	0.0059	0.0728	0.4697	0.6800
0.4196	0.2521	0.6300	0.7338	0.0154	0.1095	0.5650	0.7338
rmsd%=9.49								
315.2 K							
0.9877	0.9823	0.0123	0.0177	0.0003	0.0002	0.0075	0.0083
0.9335	0.9620	0.0665	0.0380	0.0004	0.0003	0.0195	0.0175
0.8600	0.9382	0.1368	0.0618	0.0004	0.0005	0.0194	0.0192
0.7804	0.8771	0.2158	0.1229	0.0004	0.0007	0.0192	0.0250
0.7081	0.8361	0.2884	0.1639	0.0005	0.0009	0.0296	0.0379
0.6183	0.7496	0.3779	0.2502	0.0004	0.0009	0.0227	0.0559
0.5234	0.5287	0.4727	0.4710	0.0010	0.0020	0.0571	0.1200
0.4188	0.4187	0.5678	0.5810	0.0406	0.0208	0.6302	0.6400
0.3069	0.3097	0.6660	0.6900	0.0797	0.0710	0.7478	0.7420
rmsd%=5.73								

The goodness of the fit, between the observed and calculated mole fractions, was calculated in terms of the root mean square deviation (rmsd). The rmsd values were calculated according to the equation of percentage root mean square deviations (rmsd%):

$$rmsd\% = 100\sqrt{\frac{\sum_{k=1}^{n}\sum_{j=1}^{2}\sum_{i=1}^{3}\left(\hat{x}_{ijk}-x_{ijk}\right)^{2}}{6n}} \qquad (15)$$

where n is the number of tie-lines, x_{exp} indicates the experimental mole fraction, x_{calc} is the calculated mole fraction, and the subscript i indexes components, j phases and $k=1,2,...n$ *(tie lines)*. The average (rmsd%) between the observed and calculated mole fractions with a reasonable error was 8.14%

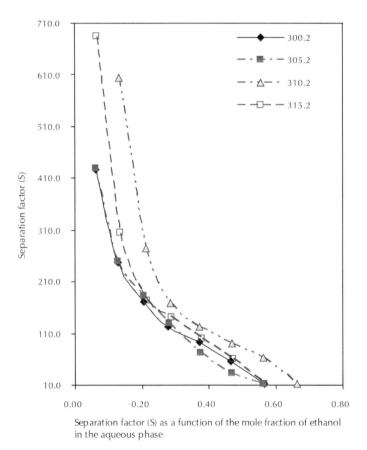

FIGURE 11.5 Factor separation (S) of formic acid as a function of the mole fraction of formic acid in aqueous phase.

The effectiveness of extraction of formic acid by cumene is given by its separation factor (S) which is an indication of the ability of cumene to separate formic acid from water. This quantity is greater than one (separation factors varying between 11.9895 and 684.3327) for the system reported here, which mean that extraction of formic acid by cumene is possible. It is, however, not constant over the whole two-phase region.

$$\ln\left(\frac{1-x_{33}}{x_{33}}\right) = A + B\ln(\frac{1-x_{11}}{x_{11}}) \quad (16)$$

The reliability of experimental measured tie-line data is determined by making an Othmer and Tobias correlation (Eq. 16) for the ternary system at each temperature. The linearity of the plots in Figure 11.7 indicates the degree of consistency of the related data (Othmer and Tobias, 1942). The Othmer and Tobias plots at different temperatures are shown in Figure 11.7 and the correlation parameters are given in Table 11.4.

TABLE 11.4 OthmerTobias equation constants for the water-formic acid–cumene ternary system.

	Othmer and Tobias Correlation		
T/K	A	B	R^2
300.2	1.9549	0.8877	0.9931
3052	1.5457	0.9895	0.9961
310.2	1.3858	0.8023	0.9949
3152	1.0264	1.2824	0.9983

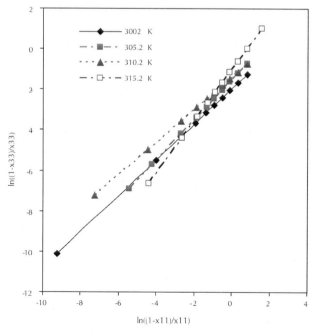

FIGURE 11.6 Othmer Tobias of the (water + formic acid +cumene) ternary system at different temperatures.

11.4 CONCLUSION

The LLE data of the ternary system composed of water + formic acid+ cumene were measured at different temperatures of 298.15, 303.15, and 308.15. The equilibrium data of the ternary mixture were also predicted by the UNIFAC method. It was found that UNIFAC with original group interaction parameters developed for LLE did not provide good prediction. The average RMSD value between the observed and calculated mole fractions with a reasonable error was 8.14% for the UNIFAC model. The solubility of water in cumene increases with amounts of formic acid added to water + cumene mixture.

It can be concluded that trans decalin showing low solubilities in water may be an adequate solvent to extract formic acid from its dilute aqueous solutions.

KEYWORDS

- Cumene
- Formic acid
- Konik gas chromatography
- Liquid–liquid equilibrium

NOTE

1 Library and information centre Royal Society of Chemistry http://www.rsc.org/knovel_library.htm

REFERENCES

Aljimaz, A. S., Fandary, M. S. H., Alkandary, J. A., and Fahim, M. A. (2000). *J. Chem. Eng. Data* **45**, 301.

Arce, A., Martinez-Ageitos, J., Rodriguez, O., and Soto, A. (2001). *J. Chem. Thermodyn.* **33**, 139–146.

Arce, A., Martinez-Ageitos, J., Rodriguez, O., and Vidal, I. (1995). *Fluid Phase Equilib.* **109**, 291–297.

Bendova, M., Rehak, K., Sewry, J. D., and Radloff, S. E. (1994). *J. Chem. Eng. Data* **39**, 320–323.

Fandary, M. S. H., Aljimaz, A. S., Al-Kandary, J. A., and Fahim, M. A. (1999). *J. Chem. Eng. Data* **44**, 1129–1131.

Fredenslund, A., Gmehling, J., and Rasmussen, P. (1977). *Vapor–liquid Equilibria Using UNIFAC a Group-Contribution Method*. Elsevier, Amsterdam.

Fredenslund, A., Jones, R. L., and Prausnitz, J. M. (1975). *AICHE J.* **21**, 1086–1099.

Garcia-Flores, B. E., Galicia-Aguilar, G. and Eustaquio-Rincon, R., and Trejo, A. (2001). *Fluid Phase Equilib.* **185**, 275–293.

Garcia, I. G., Perez, A. Ch., and Calero, F. C. (1988). *J. Chem. Eng. Data* **33**, 468–472.

Ghanadzadeh, H. and Ghanadzadeh, A. (2003). *J. Chem. Thermodyn.* **35**, 1393–1401.

Ince, E. and Ismail Kirbaslar, S. (2003). *J. Chem. Thermodyn.* **35**, 1671–1679.

Jassal, D. S., Zhang, Z., and Hill, G. A. (1994). *Can. J. Chem. Eng.* **72**, 822–826.

Kollerup, F. and Daugulis, A. J. (1985). *Can. J. Chem. Eng.* **63**, 919–927.

Magnussen, T., Rasmussen, P., and Fredenslund, A. (1981). *Ind. Eng. Chem. Process Des. Dev.* **20**, 331–339.

Othmer, D. F. and Tobias, P. E. (1942). *Ind. Eng. Chem.* **34**, 693–700.

Peschke, N., and Sander, S. I. (1995). *J. Chem. Eng. Data* **40**, 315–320.

Sola, C., Casas, C., Godia, F., Poch, M., and Serra, A. (1986). *Biotechnol. Bioeng. Symp.* **17**, 519–523.

Weast, R. C. (1989). *Handbook of Chemistry and Physics, 70th ed.* CRC Press, Boca Raton, Florida.

Zhang, S., Hiaki, T., Hongo, M., Kojima, K. (1998). *Fluid Phase Equilib.* **144**, 97–112.

Zhang, Z. and Hill, G. A. (1991). *J. Chem. Eng. Data* **36**, 453–456.

12 Molecular Thermodynamics Process Control in Fluid-phase Equilibria

CONTENTS

12.1 Introduction 116
12.2 Equation of State; Theoretical Consideration 117
    12.2.1 General Form of Equation 117
    12.2.2 A Simple Theoretical Model for the Covolume Term 120
12.3 Conclusion 122
12.4 Acknowledgment 122
Keywords 123
References 123

NOMENCLATURES

a	=	Radius of the spherical core in Kihara's model
b	=	Covolume
B	=	Second virial coefficient
c	=	Parameter of EOS
C	=	Third virial coefficient
D	=	Forth virial coefficient
k	=	Boltzman's constant
P	=	Pressure
r	=	Intermolecular distance
R	=	Universal gas constant and range of SW potential well
T	=	Temperature
U	=	Internal energy
v	=	Volume
Z	=	Compressibility factor

Greek Letter

α	=	Equation attractive parameter

ϕ	=	Fraction of collisions
λ	=	Constant of equation
ρ	=	Density
σ	=	Collision diameter
ε	=	Depth of the energy well of SW and Kihara's potential, and relative translational energy in the center line direction
Γ	=	Potential function

Subscripts

∞	=	Infinite temperature
ave	=	Average
HS	=	Hard sphere
SW	=	Square well
rep	=	Repulsion
r	=	Reduced property

Abbreviation

EOS	=	Equation of state
SRK	=	EOS of Soave Redlich Kwong
PR	=	EOS of Peng Robinson
A	=	Avogadro
K	=	Kihara's model

12.1 INTRODUCTION

The accurate prediction of thermodynamic properties of natural gas systems is of interest for gas industry. Compressibility factors are used in energy and flow metering. It is also used in calculations of gas pressure gradient in tubing and pipelines. When large volumes of gas are traded between producers, distributors, and consumers, error in the estimation of the amount of gas involved are of real economic significance. In gas condensate reservoirs, well-productivity often declines rapidly when pressure drops below the dew point pressure near-wellbore. Therefore, it is very important to accurately determine the dew point pressure. The pressure and temperature of most natural gas mixtures can be found up to 150 MPa and 500 K, respectively (Nasrifar and Boland, 2006). At these conditions, methane, ethane, and nitrogen are almost always supercritical while other hydrocarbons are subcritical. Thus, the equation of state of natural gas mixture must be accurate at supercritical and subcritical behavior of methane and heavy hydrocarbons, respectively.

Cubic equations of state are commonly used in the gas industry to predict phase behavior and volumetric properties of hydrocarbon reservoir fluids. In general, these equations follow the physical fact suggested by first-order perturbation theory according to which pressure can be expressed in terms of additive and repulsive effects (Reed and Gubbins, 1973).

The equation of state developed so far consisting repulsive and attractive term, are mostly empirical (Valderrama, 2003; Wei, and Sadus, 2000) and only a few of them are semi-empirical (Lee et al. 1987; Nasrifar and Boland, 2004; Nasrifar and Moshfeghian, 2004). Equation of state parameters have been generally determined and correlated by matching measured and predicted properties of pure substances. In this chapter, we drive an equation of state based on perturbation theory consist of a hard sphere as repulsive contribution and an attractive term based molecular thermodynamics. The attractive term and a new derived function for its parameter, α, are based on a theoretical approach.

12.2 EQUATION OF STATE; THEORETICAL CONSIDERATION

12.2.1 General form of Equation

The perturbation theory is an attractive alternative for modeling the compressibility factor. This theory has the advantage of having a sound thermodynamic basis. In the perturbation theory, the total compressibility factor is divided into a reference part and a perturbed part such as:

$$Z = Z_{rep} + Z_{att} \qquad (1)$$

or in an equivalent form

$$P = P_{rep} + P_{att} \qquad (2)$$

As a reference point, the repulsive term is usually taken the hard sphere compressibility factor (pressure). In our derivation we use the term proposed by Edalat and Hajipour (2006) to account for the repulsive effects

$$Z_{HS} = \frac{v + 0.852b}{v - 0.401b} \qquad (3)$$

$$P_{rep} = \frac{RT(v + 0.852b)}{v(v - 0.401b)} \qquad (4)$$

This term is capable to predict the compressibility factor of hard sphere fluid with packing fraction up to 0.497 with a satisfactory accuracy. If parameter b is assumed to be independent of temperature, the following exact thermodynamic relation is held.

$$\Delta U = \int_v^\infty -T^2 \left(\frac{\partial}{\partial T} \left(\frac{P}{T} \right) \right)_v dv \qquad (5)$$

we have

$$\left(\frac{\partial \Delta U}{\partial v}\right)_T = -P_{att} + T\frac{\partial P_{att}}{\partial T} \qquad (6)$$

in which

$$\Delta U = U - U_\circ \qquad (7)$$

where $U_\circ$ is the internal energy of completely separated molecules at the temperature in question. Although, in the well known equations of state such as SRK and PR equations, the co-volume, b, is independent of temperature, it should be noted that this assumption may causes considerable errors for predicting thermodynamic properties at very high pressure, especially in supercritical region.

In this study, we assume a temperature dependant covolume and using equation (5) the results will be;

$$\left(\frac{\partial \Delta U}{\partial v}\right)_T = \frac{1.253RT^2 b'}{(v-0.401)^2} - P_{att} + T\frac{\partial P_{att}}{\partial T} \qquad (8)$$

where

$$b' = \frac{\partial b(T)}{\partial T} \qquad (9)$$

On the other hand, we would like to seek a relation between hard sphere equation in the virial form of equation of state. The virial form of EOS in pressure explicit form can be written as;

$$Z = 1 + \frac{B}{v} + \frac{C}{v^2} + \frac{D}{v^3} + \cdots \qquad (10)$$

The second virial coefficient is related to the potential functions as following (McQuarrie, 1973).

$$B = 2\pi N_A \int_0^\infty \left\{1 - \exp\left(\frac{-\Gamma(r)}{kT}\right)\right\} r^2 dr \qquad (11)$$

where $\Gamma(r)$ is the potential energy. In order to determine B, we use the square-well (SW) potential. Although, this potential is obviously unrealistic, its mathematical simplicity and flexibility make it useful for practical calculations. The flexibility arises from the SW potential's three adjustable parameters (Prausnitz et al., 1999). It is expressed by McQuarrie (1973)

$$\Gamma_{SW}(r) = \begin{cases} \infty & r < \sigma_{SW} \\ -\varepsilon_{SW} & \sigma_{SW} < r < R_{SW}\sigma_{SW} \\ 0 & R_{SW}\sigma_{SW} < r \end{cases} \qquad (12)$$

where $\Gamma_{SW}(r)$ is the potential energy, σ_{SW} is the size parameter, R_{SW} is the range of potential well and ε_{SW} is the depth of the potential well. Inserting equation (10) into equation (9) yields

$$B_{SW} = b_{SW}\left[1-\left(R_{SW}^{3}-1\right)\left(\exp\left(\varepsilon_{SW}/kT\right)-1\right)\right] \quad (13)$$

where b_{SW} is equal to $2\pi N_A \sigma_{SW}^3 / 3$.

The third and higher terms in virial series is replaced with a term which decreases deviation of truncated form. The replacing term is a function of temperature and volume as:

$$-\frac{g(v)}{RT} \quad (14)$$

Since, real molecules at infinite temperature behavior is the same as hard-body fluids, we have

$$\lim_{T\to\infty}\left(1+\frac{B_{SW}}{v}-\frac{g(v)}{RT}\right)=\frac{v+0.852b_{\infty}}{v-0.401b_{\infty}} \quad (15)$$

in which

$$b_{\infty} = \lim_{T\to\infty} b(T) \quad (16)$$

This constrain can not be met with the first and second terms of equation (10). To seek is defined as following form

$$\frac{\lambda B_{SW}}{v+c} \quad (17)$$

where

$$\lambda = 1.253 \quad (18)$$

$$c = -0.401b_{\infty} \quad (19)$$

Inserting this truncated virial series into equation (5) yields;

$$\left(\frac{\partial \Delta U}{\partial v}\right)_T = -\frac{\lambda b_{SW} R a_1 a_2}{v(v+c)}\exp\left(\frac{a_2}{T}\right)+f(v) \quad (20)$$

where

$$f(v) = \frac{g(v)}{v} \quad (21)$$

$$a_1 = -\left(R_{SW}^3 - 1\right) \quad (22)$$

$$a_2 = \varepsilon_{SW}/k \qquad (23)$$

Comparing equations (8) and (20), and then solving the differential equation obtained, the attractive term of equation (6) will be as following form.

$$P_{att} = T\left(\frac{\lambda b_{SW} Ra_1}{v(v+c)}\exp\left(\frac{a_2}{T}\right) - \frac{1.253R}{0.401(v-0.401b)} + c_2\right) - f(v) \qquad (24)$$

where c_2 is the integration constant.

If the P_{att} should monotonically decrease to zero with temperature, one can conclude that

$$f(v) = \frac{\lambda b_{SW} Ra_1 a_2}{v(v+c)} \qquad (25)$$

$$c_2 = -\frac{\lambda b_{SW} Ra_1}{v(v+c)} + \frac{1.253R}{0.401(v-0.401b_\infty)} \qquad (26)$$

By substituting equations (22) and (23) into the equation (2), the new equation of state will be;

$$P = \frac{RT(v+0.852b)}{v(v-0.401b)} - \frac{\alpha(T_r)}{v(v+c)} + \frac{1.253RT(c-b)}{(v-0.401b)(v+c)} \qquad (27)$$

where

$$\alpha(T_r) = b_{SW} Ra_1 \lambda T_c T_r \left(1-\exp\left(\frac{a_2/T_c}{T_r}\right)\right) + b_{SW} Ra_1 a_2 \lambda \qquad (28)$$

12.2.2 A Simple Theoretical Model for the Covolume Term

If molecules are considered to behave by Kihara's model (Kihara, 1953), the following form of potential function can be applied

$$\Gamma = \begin{cases} \infty & r < 2a \\ 4\varepsilon_K\left[\left(\frac{\sigma_K - 2a}{r-2a}\right)^{12} - \left(\frac{\sigma_K - 2a}{r-2a}\right)^{12}\right] & r > 2a \end{cases} \qquad (29)$$

where a is radius of the spherical core, ε_K is the depth of the energy well, σ_K is the collision diameter, and r is the intermolecular distance. Using the above potential and a distribution function representing the collision energy of molecules, one can obtain an equation which shows the functional form of the average diameter of molecules.

In this study we use the function developed by Moore and Pearson (1981)

Molecular Thermodynamics Process Control in Fluid-phase Equilibria 121

$$\phi = \exp(-\varepsilon/kT) \quad (30)$$

where ϕ is the fraction of collisions which has a value of relative translational energy in the center line direction, greater than ε. Although, this function has been obtained assuming Maxwell's distribution, it can prepare a relatively good approximation of behavior of real fluids (Morris and Present, 1969).

Using equations (29) and (30), one can obtain

$$b = b_{ave}(T) = \frac{2\pi}{3} N_A \times \left(\frac{(\sigma_K - 2a) \times 2^{1/6}}{\left(1 + \left(1 + kT/\varepsilon_K\right)^{0.5}\right)^{1/6}} + 2a \right)^3 \quad (31)$$

TABLE 12.1 Coefficient of polynomial.

Compound	T^4	T^3	T^2	T	T^0
N_2	6.1295E-18	− 1.8495E-14	2.4081E-11	− 2.1733E-8	5.6997E-5
CH_4	5.1938E-18	− 1.6461E-14	2.2553E-11	− 2.1275E-8	5.6952E-5
C_2H_6	1.0591E-18	− 3.8159E-15	6.6338E-12	− 1.7234E-8	5.4225E-5
C_4H_{10}	7.9836E-19	− 3.3310E-15	7.1138E-12	− 1.3930E-8	1.3231E-4
C_5H_{12}	5.4911E-19	− 2.4339E-15	5.7286E-12	− 1.2988E-8	1.6035E-4

which is the desired form of $b(T)$. The equation shows that b is a monotonically decreasing function of temperature. In order to simplify the above equation one can fit a polynomial function. According to our calculations, a forth order polynomial can be used resulting in a value of R square greater than 0.9999. Some results are presented in Table 12.1.

There are a few equations of state with the same trend of the variation of b as a function of temperature (Zebolsky, 2000).

It is noteworthy that in contrast to many equations of state, such as the equations with the van der Walls repulsive term, this equation can meet the following inequality as shown in figure (12.1)

$$\frac{\partial P_{real\ fluid}}{\partial \rho} < \frac{\partial P_{hard\ sphere\ fluid}}{\partial \rho} \quad (32)$$

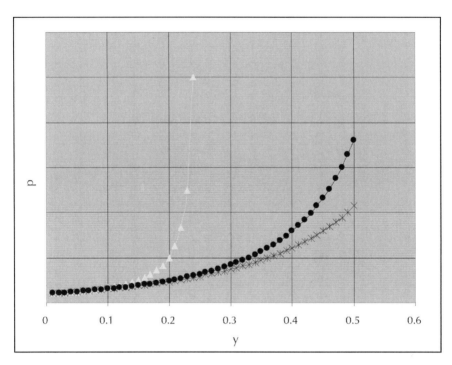

FIGURE 12.1 Predicted behavior of nitrogen at very high pressure for the derived equation in comparison with PR (Peng and Robinson, 1976) equation of state and the equation proposed by Khoshkbarchi and Vera (1997) for hard sphere fluids.

In order to predict natural gas dew point pressures, we used the van der Waals mixing rules and the decay function proposed by Pedersen et al. (1992) for the heavy ends of different natural gas mixtures containing plus fraction. Predicted dew point pressures were in good agreement with experimental data (Elsharkawy, 2002).

12.3 CONCLUSION

According to the above derivation, molecular potential parameters are clearly related to different parameters of the cubic equation of state. In contrast to many well-known EOSs such as SRK (Soave, 1972), the functional form of α derived in this work, shows the correct behavior of fluids at high temperatures (Segura et al., 2003). Since, α function is continues at T_c no anomalous behavior may be seen in predicted calorimetric properties at and near the critical point (Deiters, 1999). In addition, as a result of using a correct temperature dependent form for the covolume good agreement with experimental data at very high pressure has been obtained.

12.4 ACKNOWLEDGMENT

This study is supported by Iranian South Oil Company and Oil and Gas Center of Excellence of Chemical Engineering Department, University of Tehran.

KEYWORDS

- **Covolume**
- **Infinite temperature**
- **Polynomial function**
- **Thermodynamic**

REFERENCES

Deiters, U. K. (1999). Remarks on Publications Dealing with Equations of State. *Fluid Phase Equilib.* **161**, 205.

Edalat, M., Hajipour, S., and Fasih, M. (2006). *A new equation of state predicts phase behavior of hydrocarbon and hydrocarbon mixtures of Iranian oil reservoirs*, 11th APCChE congress, Kuala Lumpur, paper ID: 548.

Elsharkawy, A. M. (2002). Predicting the dew point pressure for gas condensate reservoirs: Empirical models and equations of states. *Fluid Phase Equilib.* **193**, 147.

Khoshbarchi, M. K. and Vera, J. H. (1997). A simplified hard sphere equation of state meeting the high and low density limits. *Fluid Phase Eqilib.* **130**, 189.

Kihara, T. (1953). Virial Coefficients and Models of Molecules in Gases. *J. Chem. Phys.* **25**, 831.

Lee, K. H., Lombardo, M., and Sandler, S. I. (1987). The Generalized van der Waals Partition Function: Application to Square-Well Fluids. *Fluid Phase Equilib.* **21**, 177.

McQuarrie, D. A. (1999). *Statistical Thermodynamics*. Harper and Row, New York.

Moore, J. W. and Pearson, R. G. (1981). *Kinetics and Mechanism*, 3rd ed., Wiley Interscience.

Morris, B. M. and Present, R. D. (1969). Second-Order Perturbation of the Velocity Distribution in a Fast Gas-Phase Reaction. *J. Chem. Phys.* **51**, 4862.

Nasrifar, Kh. and Boland, O. (2004). Square-Well Potential and a New Ω α Function for the Soave-Redlich-Kwong Equation of State. *Ind. Eng. Chem. Res.* **43**, 6901.

Nasrifar, Kh. and Boland, O. (2006). Prediction of Thermodynamic Properties of Natural Gas Mixtures Using 10 Equations of State Including a New Cubic Two-Constant Equation of State. *J. Petro. Sci. Eng.* **51**, 253.

Nasrifar, Kh. and Moshfeghian, M. (2004). Application of an Improved Equation of State to Reservoir Fluids: Computation of Minimum Miscibility Pressure. *J. Petro. Sci. Eng.* **42**, 223.

Pedersen, K. S., Blilie, A. L., and Meisingset, K. K. (1992). PVT Calculations on petroleum reservoir fluids using measured and estimated compositional data for the plus fraction. *Ind. Eng. Chem. Res.* **3**, 1378.

Peng, D. Y. and Robinson, D. B. (1976). A new two constants equation of state. *Ind. Eng. Chem. Fund.* **15**, 59.

Prausnitz, J. M., Lichtenthaler, R. N., and Gomes de Azevedo, E. (1999). *Molecular thermodynamics of fluid-phase equilibria*, 3rd ed., Prentice Hall, USA.

Reed, T. and Gubbins, K. E. (1973). *Applied Statistical Mechanics*. McGraw-Hill, New York.

Segura, H., Kraska, T., Mejia, A., Wisniak, J., and Polishuk, H. (2003). Unnoticed Pitfalls of Soave-Type Alpha Function in Cubic Equation of State. *Ind. Eng. Chem. Res.* **42**, 5662.

Sherwood, A. E. and Prausnitz, J. M. (1964). Intermolecular potential function and the second and third virial coefficients. *J. Chem. Phys.* **41**, 429.

Soave, G. (1972). Equilibrium Constants from a Modified Redlich-Kwong Equation of State. *Chem. Eng. Sci.* **27**, 1197.

Valderrama, J. O. (2003). The State of Cubic Equations of State. *Ind. Eng. Chem. Res.* **42**, 1603.

Wei, Y. S. and Sadus, R. J. (2000). Equations of State For Calculation of Fluid-Phase Equilibria. *AICHE Journal* **46**, 169.

Zebolsky, D. M. (2000). Comparisons of supercritical properties from an equation of state with a hard sphere repulsive pressure term and from the Peng-Robinson equation of state. *Ind. Eng. Chem. Res.* **39**, 3521.

13 Estimation of Liquid-liquid Equilibrium Using Artificial Neural Networks

CONTENTS

13.1 Introduction 126
13.2 Group Method of Data Handling 127
       13.2.1 GMDH-Type Neural Network 127
13.3 Estimation of LLE Using the GMDH Models 128
13.4 Conclusion 132
Keywords 133
References 133

NOMENCLATURES

C = Total number of components
calc = Calculated concentration superscript
exp = Experimental concentration superscript
f = Actual function
F = Two-variable quadratic function
I = Oil phase superscript
II = Alcoholic phase superscript
D_c = C th distribution coefficient
M = Number of observation of input–output data pairs
m = Mole fraction of components
n = Number of input variables
P = Total number of phases
t = Tie-line subscript
T = Total number of tie-lines
x = Input of polynomial functional node
X = Input vector
y = Actual output of polynomial functional node
$\hat{y}$ = Predicted output of polynomial functional node

13.1 INTRODUCTION

Modeling and prediction of phase equilibrium data are very important for simulation, design, and optimize of separation operations. A large amount of investigation has been carried out in recent years on the liquid–liquid equilibrium (LLE) measurements of ternary systems, in order to understand and provide further information about the phase behavior of such systems. Usually, the thermodynamic models have been successfully applied for the correlation of several LLE systems but these conventional methods for LLE data prediction of complex systems are tedious and involve a certain amount of empiricism. To avoid these limitations, new correlation methods were developed by using artificial neural networks (ANNs) which have been recently applied to many prediction tasks. The ANNs are non linear and highly flexible models that used to model complex non-linear relationships. The ANNs offer the potential to overcome the limitations of the traditional thermodynamic models and polynomial correlation method for the complicated systems, especially in estimating the LLE and vapor–liquid equilibrium (VLE) (Ganguly, 2003; Guimaraes and McGreavy, 1995; Mohanty, 2005; Sharma et al., 1999, Urata et al., 2002). The ANNs are able to determine the relationship between a set of input data and the corresponding output data without the need for predefined mathematical equations between these data but the inherent complexity in the design of neural networks in terms of understanding the most appropriate topology and coefficients has a great impact on their performance (Nariman-zadeh and Jamali, 2007). Conversely, the group method of data handling (GMDH) (Ivakhnenko, 1971) is aimed to identify the functional structure of a model hidden in the empirical data. The GMDH creates adaptively models from data in form of networks of optimized transfer functions in a repetitive generation of layers. Neither, the number of nodes (neurons) and layers in the network, nor the transfer functions of neurons are predefined. All these are adjusted during the process of self-organization by the process itself. As a result, an explicit analytical model representing relevant relationships between input and output variables is available immediately after modeling (Onwubolu, 2007). The GMDH uses a feed-forward network structure based on short-term polynomial transfer function whose coefficients are obtained using regression technique (Farlow, 1984). The GMDH was developed for complex systems modeling, prediction, identification and approximation of multivariate processes, diagnostics, pattern recognition, and clusterization of data sample. It was proved that for inaccurate, noisy, or small, data can be found best optimal simplified model, accuracy of which is higher and structure is simpler than structure of usual full physical model. In this work, an LLE prediction method was developed by using GMDH algorithm to predict LLE data of a ternary system (water + 1-propanol+ diisopropyl). Using existing data from (Ghanadzadeh and Haghi, 2006), the proposed model was trained and then used to predicting of LLE data in aqueous and organic phases. Then, the predicted data from the GMDH model compared with the experimental data. Also, mean deviations obtained by UNIFAC and proposed model have been compared. The phase diagrams for the studied ternary system including both the experimental and predicted tie-lines are presented.

13.2 GROUP METHOD OF DATA HANDLING

The GMDH algorithm is a self-organizing approach by which gradually complicated models are generated based on the evaluation of their performances on a set of multi-input–single-output data pairs. The GMDH was first proposed by Alexy G. Ivakhnenko (Ivakhnenko, 1971) as a multivariate analysis method for complex systems modeling and identification. The GMDH can be used to model complex systems without having specific knowledge of the systems. Therefore, main idea of GMDH is to obtain a mathematical model of the object under study.

The classical GMDH method (Farlow, 1984; Ivakhnenko, 1971) is based on an underlying assumption that the data can be modeled by using an approximation of the Volterra Series or Kolmogorov-Gabor polynomial (Madala and Ivakhnenko, 1994) as shown in the following equation:

$$y = a_0 + \sum_{i=1}^{M} a_i x_i + \sum_{i=1}^{M}\sum_{j=1}^{M} a_{ij} x_i x_j + \sum_{i=1}^{M}\sum_{j=1}^{M}\sum_{k=1}^{M} a_{ijk} x_i x_j x_k \ldots \qquad (1)$$

Ivakhnenko accomplished this by using a feed-forward self-organizing polynomial functional network.

13.2.1 GMDH-type Neural Network

Traditional GMDH is a feed-forward type neural network formed by neurons whose transfer function is a short-term polynomial of two variables, equation (2). Each layer of network consists of nodes generated to take a specific pair of the combination of inputs as its source. Each node produces a set of coefficients a_i where $i \in \{0, 1, 2, \ldots, 5\}$ such that equation (2) is estimated using the set of *training* data. This equation is tested for fit by determining the mean square error of the predicted $\hat{y}$ and actual y values as shown in equation (3) using the set of *testing* data.

$$\hat{y}_n = a_0 + a_1 x_{i_n} + a_2 x_{j_n} + a_3 x_{i_n} x_{j_n} + a_4 x_{i_n}^2 + a_5 x_{j_n}^2 \qquad (2)$$

$$e = \sum_{n=1}^{N} \left(\hat{y}_n - y_n\right)^2 \qquad (3)$$

In determining the values of a that would produce the "best fit", the partial derivatives of equation (3) are taken with respect to each constant value a_i and set equal to zero.

$$\frac{\partial e}{\partial a_i} = 0 \qquad (4)$$

Expanding equation (4) results in the following matrix equations that are solved using the *training* dataset:

$$A = Y^T Y \qquad (5)$$

where Y and A are

$$Y = \begin{pmatrix} 1 & x_i & x_j & x_i x_j & x_i^2 & x_j^2 \end{pmatrix} \tag{6}$$

$$A = \begin{pmatrix} 1 & x_i & x_j & x_i x_j & x_i^2 & x_j^2 \\ x_i & x_i^2 & x_i x_j & x_i^2 x_j & x_i^3 & x_i x_j^2 \\ x_j & x_i x_j & x_j^2 & x_i x_j^2 & x_i^2 x_j & x_j^3 \\ x_i x_j & x_i^2 x_j & x_i x_j^2 & x_i^2 x_j^2 & x_i^3 x_j & x_i x_j^3 \\ x_i^2 & x_i^3 & x_i^2 x_j & x_i^3 x_j & x_i^4 & x_i^2 x_j^2 \\ x_j^2 & x_i x_j^2 & x_j^3 & x_i x_j^3 & x_i^2 x_j^2 & x_j^4 \end{pmatrix} \tag{7}$$

$$x = \begin{pmatrix} a_0 & a_1 & a_2 & a_3 & a_4 & a_5 \end{pmatrix} \tag{8}$$

$$b = (yY)^T \tag{9}$$

This system of equations then can be written as:

$$\sum_{n=1}^{N} Ax = \sum_{n=1}^{N} b \tag{10}$$

The set of coefficients (x) are computed by solving the system using the *training* set of data. Using these coefficients in equation (2), the error of node is computed by processing the set of *testing* data in equations (2) and (3). The *testing* data error is then used as a measure of nodal fitness. A GMDH layer sorts its nodes based on the nodal fitness produced, saving the best **N** nodes. The output of each node (y_n value) will be used as an input in the next layer. This process is continued as long as each subsequent layer$_{(n+1)}$ produces a better result than layer$_{(n)}$. When layer$_{(n+1)}$ is found to not be as good as layer$_{(n)}$, the process is halted. In Figure 13.1, the structure of the feed-forward GMDH-type neural network is shown.

13.3 ESTIMATION OF LLE USING THE GMDH MODELS

In this study, four GMDH models were designed using existing experimental dataset (Ghanadzadeh, 2008). The dataset consisted of 36 points at four different temperatures (300.2, 305.2, 310.2, and 315.2 K). There were nine experimental data points at each temperature separately. Using 36 points dataset, we proposed a GMDH model. The 36-point dataset divided in two parts, namely, *training* and *testing* sets. The *training* and *testing* sets consisted of 75% and 25% of data points, respectively. Each point in *training* and *testing* sets consisted of nine mole fraction values. Three mole fractions of the feed components used as inputs of the GMDH-type neural network and other six mole fractions in aqueous and organic phases used as desired outputs of the network. We applied the input–output data pairs to the GMDH-type neural networks in order to obtain a model which can be used to predict of mole fractions in aqueous and organic phases at different temperatures. The GMDH model consisted of six polynomial equations that were used to calculate the LLE data. The experimental points and predicted

Estimation of Liquid-liquid Equilibrium Using Artificial Neural Networks 129

tie-lines from GMDH model for the system water + 1-propanol + diisopropyl at 300.- –315.2 K plotted in Figures 13.5. These figures indicate that GMDH model provided a good estimation of the LLE data in both phases.

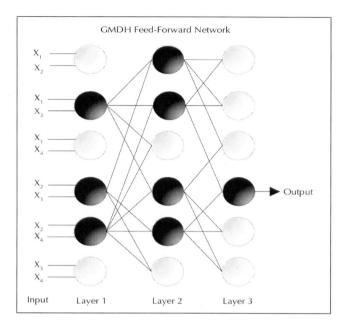

FIGURE 13.1 A GMDH-type neural network.

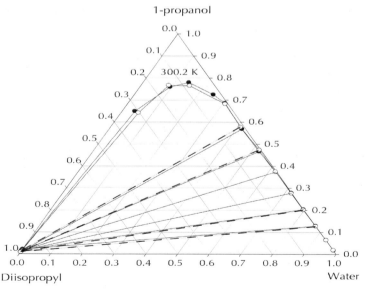

FIGURE 13.2 Prediction of the experimental data for (water + 1-propanol+ diisopropyl) system at 300.2 K. (●) Experimental points; (○) calculated points.

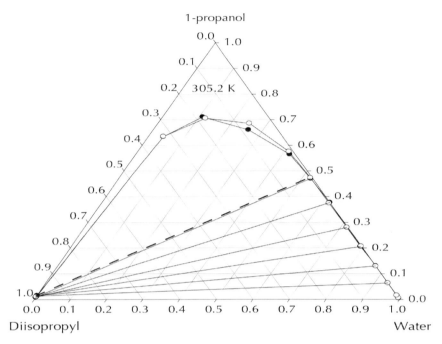

FIGURE 13.3 Prediction of the experimental data for (water + 1-propanol+ diisopropyl) system at 305.2 K. (●) Experimental points; (○) calculated points.

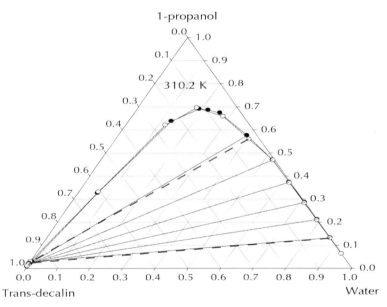

FIGURE 13.4 Prediction of the experimental data for (water + 1-propanol+ diisopropyl) system at 310.2 K. (●) Experimental points; (○) calculated points.

Estimation of Liquid-liquid Equilibrium Using Artificial Neural Networks 131

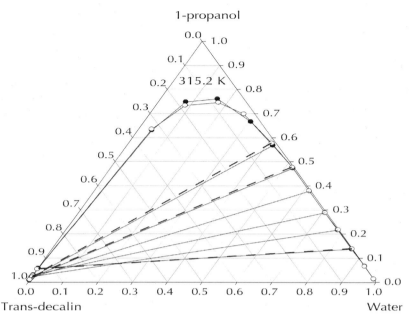

FIGURE 13.5 Prediction of the experimental data for (water + 1-propanol+ diisopropyl) system at 315.2 K. (●) Experimental points; (○) calculated points.

We used the GMDH model to estimate the distribution coefficient. The distribution coefficient of 1-propanol can be calculated by equation (11).

$$D_2 = \frac{m_{23}}{m_{21}} \qquad (11)$$

where m_{23} are the mole fractions of 1-propanol in organic phase and m_{21} are the mole fractions of 1-propanol in aqueous phase, respectively. The experimental and estimated distribution coefficient of 1-propanol between the phases as a function of the mole fraction of 1-propanol in aqueous phase for the ternary system is presented in Figure 13.6. This diagram is plotted for one temperature (300.2 K) in order to reduce confusion of them.

The comparisons between the experimental and calculated composition of each component of two phases were made through root mean square deviation (rmsd). The rmsd were calculated from the difference between the experimental data and predictions of the GMDH model at different temperatures according to the following equation:

$$\text{rmsd} = \sqrt{\frac{\sum_t^T \sum_p^P \sum_c^C \left(m_{cpt}^{\text{exp}} - m_{cpt}^{\text{calc}}\right)^2}{6T}} \qquad (12)$$

where T is the total number of tie-lines, C is the total number of components and P is number of phases. m is the mole fraction, the subscripts c, p and t are component, phase and tie-line, respectively; exp and calc refer to experimental and calculated mole fractions. In Table 13.2, the calculated deviations are compared with the rmsd values reported by Ghanadzadeh (2008). As be shown, the GMDH method provides better estimation accuracy than the UNIFAC model.

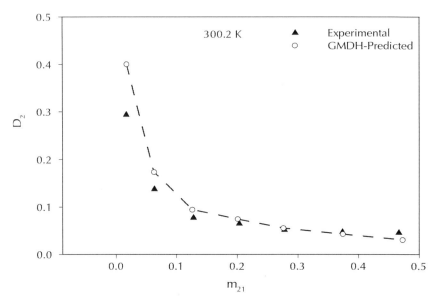

FIGURE 13.6 Distribution coefficient (D2) of 1-propanol as a function of the mole fraction (m21) of 1-propanol in aqueous phase.

TABLE 13.2 Root mean square deviations in different models.

System	GMDH	UNIQUAC (Ghanadzadeh, 2008)
Water+1-propanol+diisopropyl	%0.55	%1.14

13.4 CONCLUSION

In this study, a GMDH model was designed using the experimental LLE data for system water + 1-propanol+ diisopropyl at 300.2–315.2 K (Ghanadzadeh and Haghi, 2006). The LLE data were predicted by the GMDH models and then compared with the experimental data. Despite the complexity of the studied system, the GMDH model allowed a good estimate accuracy to be achieved for phase equilibrium. Also, the mean deviation of the proposed GMDH models was lower than 0.55% in relation to the experimental data. The GMDH model provides a better estimation compared with the UNIQUAC model. The GMDH method may be suitable to use in place of ANN and conventional methods to estimate LLE.

KEYWORDS

- Aqueous phase
- Clusterization
- Empiricism
- Neurons
- Ternary systems

REFERENCES

Dayhoff, J. E. (1990). *Neural Network Architectures*. Van Nostrand Reinhold, New York.

Farlow, S. J. (1984). *Self-Organizing Method in Modeling: GMDH Type Algorithm*. Dekker, New York.

Ganguly, S. (2003). Prediction of VLE data using radial basis function network. *Comput. Chem. Eng.* **27**, 1445–1454.

Ghanadzadeh, H. (2006). Liquid–liquid equilibrium data for water + 1-propanol + diisopropyl: Measurement and predication. *Fluid Phase Equilibria* **243**, 45–50.

Guimaraes, P. R. B. and McGreavy, C. (1995). Flow of information through an artificial neural network. *Comput. Chem. Eng.* **19**(S1), 741–746.

Ivakhnenko, A.G. (1971). Polynomial theory of complex systems. *IEEE Trans. Syst. Man Cybern* **1**, 364–378.

Krose, B. and Van der Smagt, P. (1996). *An Introduction to Neural Networks*. University of Amesterdam, The Netherlands.

Madala, H. R. and Ivakhnenko, A. G. (1994). *Inductive Learning Algorithms for Complex Systems Modeling*, CRC Press Inc., Boca Raton.

Mohanty, S. (2005). Estimation of vapour liquid equilibria of binary systems, carbon dioxide–ethyl caprylate and ethyl caprate using artificial neural networks. *Fluid Phase Equilib* **235**, 92–98.

Nariman-Zadeh, N., Darvizeh, A., Felezi, M. E., and Gharababaei, H. (2002). Polynomial modelling of explosive compaction process of metallic powders using GMDH-type neural networks and singular value decomposition. *Modelling Simul. Mater. Sci. Eng.* **10**, 727–744.

Nariman-Zadeh, N. and Jamali, A. (2007). Pareto Genetic Design of GMDH-type Neural Networks for Nonlinear Systems. *Proc. International Workshop on Inductive Modelling—IWIM2007* 96–103.

Onwubolu, G. C. (2007). Data Mining using Inductive Modelling Approach. *Proc. International Workshop on Inductive Modelling—IWIM2007* 78–86.

Sharma, R., Singhal, D., Ghosh, R., and Dwivedi, A. (1999). Potential applications of artificial neural networks to thermodynamics: Vapor–liquid equilibrium predictions. *Comput. Chem. Eng.* **23**, 385–390.

Urata, S., Takada, A., Murata, J., Hiaki, T., and Sekiya, A. (2002). Prediction of vapor–liquid equilibrium for binary systems containing HFEs by using artificial neural network. *Fluid Phase Equilib* **199**, 63–78.

14 Optimization of Process Control of Water + Propanoic Acid+ 1-Octanol System

CONTENTS

14.1 Introduction 135
14.2 Experimental..136
    14.2.1 Materials 136
    14.2.2 Apparatus and Procedure 136
    14.2.3 Uniquac Model 137
14.3 Discussion and Result..137
    14.3.1 LLE Measurment 137
14.4 Simulation and Separation Program 143
14.5 Conclusion...144
Keywords 145
References...145

14.1 INTRODUCTION

Over the last 3 decades, there has been a resurgence of interest in large-scale production of fermentation chemicals due to the sharp increase in petroleum cost. So, the potential role of a new energy efficient fermentation technology is receiving growing attention. The current economic impact of fermentation chemicals, however, is still limited, in large part because of difficulties of product recovery. Thus, for fermentation products to penetrate the organic chemicals industry, substantial improvements in the existing recovery technology are needed. Organic acids are widely used in the food, pharmaceutical, and chemical industries. Fermentation technology for the production of organic acids in particular has been known for more than a century and acids have been produced in aqueous solutions. Propionic acid is used in the manufacture of herbicides, chemical intermediates, artificial fruit flavors, pharmaceuticals, cellulose acetate propionate, and preservatives for food, animal feed, and grain (Playne, 1985). Commercial production of propionic acid is chiefly carried out by chemical synthesis from petroleum feedstocks (Playne, 1985), but fermentation is an attractive alternative to produce this acid from renewable resources. Several carbon sources have been used for this fermentation such as glucose (Emde and Schink, 1990), xylose (Carrondo et al.,

1988), maltose (Babuchowski et al., 1993), sucrose (Quesada-Chanto et al., 1994) and whey lac tose (Lewis and Yang, 1992). The conventional method for the recovery of propionic acid from fermentation broth is the calcium hydroxide precipitation method. This method of recovery is expensive and unfriendly to the environment as it consumes lime and sulfuric acid and also produces a large quantity of calcium sulfate sludge as solid waste. Thus, there is a need to look at other methods of producing propionic acid.

Reactive extraction with specified extractant giving a higher distribution coefficient has been proposed as a promising technique for the recovery of carboxylic and hydroxycarboxylic acids. This method is advantageous for alcohol and organic fermentations (Wasewar et al., 2003). Some of the advantages include increased reactor productivity, ease in reactor pH control without requiring base addition, and use of a high-concentration substrate as the process feed to reduce process wastes and production costs. This method may also allow the process to produce and recover the fermentation product in one continuous step and reduce the down stream processing load and the recovery costs.

Long-chain, aliphatic amines are effective extractants for separation of carboxylic acids from dilute aqueous solution (Yang et al., 1991). Generally, the amine extractants are dissolved in a diluent, an organic solvent that dilutes the extractant. It controls the viscosity and density of the solvent phase. In order to improve the amine's solvation power, diluents such as oleyl alcohol, chloroform, methyl isobutyl ketone, and 1-octanol have been used. The diluents affect the basicity of the amine, the stability of the acid amine complex formed and its solvation power. The pH of the aqueous phase is an important parameter for the reactive extraction of organic acids (Kahya et al., 2001). In the present study, various pure diluents are used for extraction of propionic acid from aqueous solution. On the basis of distribution coefficients, reactive extraction is also carried out with amine extractant for the recovery of propionic acid.

14.2 EXPERIMENTAL

14.2.1 Materials

The chemicals propionic acid (99.9%) and 1-octanol (99.9%) were obtained from Merck and were used without further purification. The purity of these materials were checked by gas chromatography. Distilled water was prepared in our laboratory and used throughout all experiments.

14.2.2 Apparatus and Procedure

The equilibrium data were determined using an experimental apparatus of a 300 ml glass cell, where the temperature of the apparatus controlled by a water jacket and maintained with an uncertainly of within ±0.001°C. The temperature was measured using a calibrated digital thermometer traceable to the NIST.The mixture was vigorously agitated by a magnetic stirrer for 4 hr. The prepared mixtures were then left to settle for 8 hr for phase separation. The samples of organic-rich phase were taken by a syringe (1 µl) from the upper layer and that of water-rich phase from a sampling tap at the bottom of the cell. Samples were analyzed using Konik gas chromatography (GC), equipped with a thermal conductivity detector (TCD) and Shimadzu C-R2AX integrator. A 2 m . 2 mm Porapak OS packed column was used to separate the components. The TCP's response was linear and calibrated with 1-octanol as in internal standard. The calibration samples were prepared by weighing with an analytical balance accurate to within ±0.0001 g. The calibration equations were used to convert the area fraction into mole fraction. Calibration coefficients were obtained by fitting a straight line to the calibration results for each composition range. The experimental error of the observed mole fraction during the calibration was about ±0.05%.

14.2.3 UNIQUAC Model

The UNIQUAC model can be used in predicting activity coefficients, y. At LLE, the activities of the component i on both phases (extracted phase and raffinate phase) are equal and the mole fractions X_i^E, x_i^R of LLE phases can be determined using the following equations:

$$(\gamma_i X_i)^E = (\gamma_i X_i)^R$$

$$\sum X_i^1 = \sum X_i^2 = 1$$

where γ_i^E and γ_i^R are the corresponding activity coefficients of component i in extracted phase and raffinate phase. Gibbs free energy can be determined using the following equations:

$$\frac{g^k}{RT} = \sum_{i=1}^{c} x_i \ln\left(\frac{\phi_i}{x_i}\right) + \frac{z}{2}\sum_{i=1}^{c} q_i x_i \ln\left(\frac{\theta_i}{\phi_i}\right) - \sum_{i=1}^{c} q_i x_i \ln\left(\sum_{j=1}^{c} \theta_j \tau \mu\right)$$

$$\ln \gamma_i = \ln \gamma_i^c + \ln \gamma_i^R$$

$$\ln \gamma_i^c = \ln\left(\frac{\phi_i}{x_i}\right) + \frac{z}{2} q_i \ln\left(\frac{\theta_i}{\phi_i}\right) + \tau_i - \frac{\phi_i}{x_i}\sum_{j=1}^{c} x_j \tau_j$$

$$\ln \gamma_i^R = q_i \left[1 - \ln\left(\sum_{j=1}^{c} \theta_j \tau_{ji}\right) - \sum_{j=1}^{c}\left(\frac{\theta_j \tau_{ij}}{\sum_{k=1}^{c} \theta_k \tau_{kj}}\right)\right]$$

14.3 DISCUSSTION AND RESULT

14.3.1 LLE Measurment

The experimental and UNIQUAC LLE data for (water + propionic acid + 1-octanol) at different temperature of (293.15–303.15) K, are presented in Table 14.1. The estimated uncertainties in the mole fraction were about 0.0005. From the LLE phase diagrams (Figure 14.1–14.4), (1-octanol + water) mixture is the only pair that is partially miscible and two liquid pairs (propionic acid + water) and (propionic acid + 1-octanol) are completely miscible. The mutual solubility of 1-octanol and water is very low and therefore, the high boiling point solveni (1-octanol) can be used as a reliable organic solvent for extraction of prpionic acid from dilute aqueous solutions.

The values of the UNIQUAC parameters for LLE were estimated by Aspen.

TABLE 14.1 Experimental and predication LLE data at each temperature for [water (1) + propionic acid (2) + octanol].

Aqueous phase (raffinate) mole fraction				Organic phase (extract) mole fraction			
X1 (water)		X2 (propionic acid)		X1 (water)		X2 (propionic acid)	
Exp.	UNIQUAC and ASPEN	EXP.	UN UNIQUAC and ASPEN	EXP.	UNQUAC and ASPEN	EXP.	UNIQUAC and ASPEN
293.15 K							
0.9998	1.0000	0	0	0,1854	0,1772	0	0
0.9894	0.9592	0.0102	0.0407	0,2591	0,1848	0,0815	0,1624
0.9790	0.9288	0.0204	0.0710	0,3231	0,1883	0,1523	0,2377
0.9618	0.9036	0.0374	0.0962	0,3637	0,1912	0,2414	0,2845
0.9536	0.8814	0.0454	0.1182	0,4113	0,1937	0,2842	0,3181
0.9455	0.8713	0.0531	0.1283	0,452	0,1948	0,3106	0,3316
0.8500	0.8465	0.147	0.1529	0,55	0,1977	0,3177	0,3614
0.7390	0.8049	0.2181	0.1941	0,6	0,2023	0,3031	0,4034
--------	0.6000	--------	0.3850	-------	0,268	-------	0,524
--------	0.5000	--------	0.4654	-------	0,32	-------	0,543

Aqueous phase (raffinate) mole fraction				Organic phase (extract) mole fraction			
X1 (water)			X2 (propionic acid)			X1 (water)			X2 (propionic acid)		
UNIQUAC and ASPEN			UNIQUAC and ASPEN			UNIQUAC and ASPEN			UNIQUAC and ASPEN		
298.15 K							
0.9999			0			0.1892			0		
0.9777			0.0222			0.1925			0,1007		
0.9593			0.0406			0.1945			0,1621		
0.9433			0.0565			0.1962			0,2048		
0.9292			0.0706			0.1978			0,2369		
0.9042			0.0955			0.2008			0,2832		
0.8931			0.1067			0.2021			0,3009		
0.7101			0.2789			0.2500			0,374		
0.6001			0.3575			0.2820			0,392		
0.4913			0.4037			0.3200			0,402		
303.15 K							
0.999			0			0,1921			0		
0.9775			0.0252			0,1978			0,1113		

TABLE 14.1 *(Continued)*

Aqueous phase (raffinate) mole fraction		Organic phase (extract) mole fraction	
X1 (water)	X2 (propionic acid)	X1 (water)	X2 (propionic acid)
UNIQUAC and ASPEN	UNIQUAC and ASPEN	UNIQUAC and ASPEN	UNIQUAC and ASPEN
0.9292	0.0706	0,2032	0,2359
0.9272	0.1170	0,2043	0,2399
0.8000	0.2060	0,2395	0,408
0.7800	0.3617	0,3045	0,432
308.15 K			
0,999	0	0,1989	0
0,9777	0,0222	0,2024	0,0999
0,9593	0,0406	0,2045	0,1608
0,9434	0,0564	0,2065	0,2031
0,9292	0,0705	0,2083	0,2349
0,9043	0,0953	0,2115	0,2808
0,8827	0,1169	0,2142	0,3135
0,8631	0,1363	0,2169	0,3387
0,8451	0,1541	0,219	0,3593
0,7678	0,2283	0,227	0,3753
0,7073	0,2821	0,244	0,3914
0,5862	0,3724	0,275	0,411
0,4651	0,4291	0,3	0,425

The experimental and ASPEN-UNIQUAC LLE data for (water + propionic acid + 1-octanol) at different temperature of (293.15–303.15) K, arc presented in Table 14.1. The estimated uncertainties in the mote fraction were about 0.0005. From the LLE phase diagrams (Figures 14.1–14.4), (1-octanol + water) mixture is the only pair that is partially miscible and two liquid pairs (propionic acid + water) and (propionic acid + 1-octanol) are completely miscible. The mutual solubility of 1-octanol and water is very low and therefore, the high boiling point solveni (1-octanol) can be used as a reliable organic solvent for extraction of prpionic acid from dilute aqueous solutions.

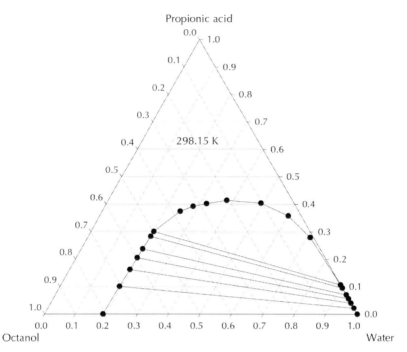

FIGURE 14.1 Prediction Correlation of the UNIQUAC data for for (water + propionic acid + octanol) system at 298.15 K (•) ASPEN and UNIQUAC points.

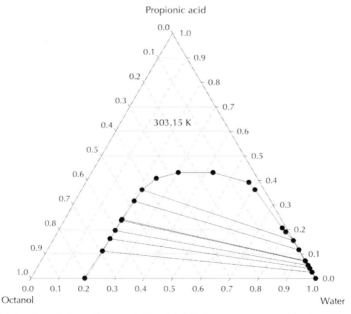

FIGURE 14.2 Correlation of the and UNIQUAC data (water + propionic acid + octanol) system at 303.15 K (•) ASPEN and UNIQUAC points.

Optimization of Process Control of Water + Propanoic Acid+ 1-Octanol System 141

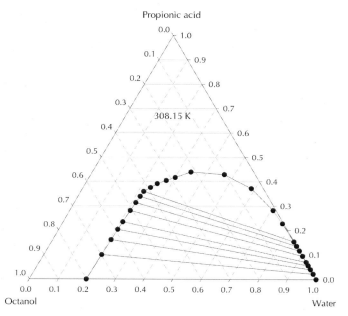

FIGURE 14.3 Correlation of the UNIQUAC data for (water + propionic acid + octanol) system at 308.15 K (•) ASPEN and UNIQUAC points.

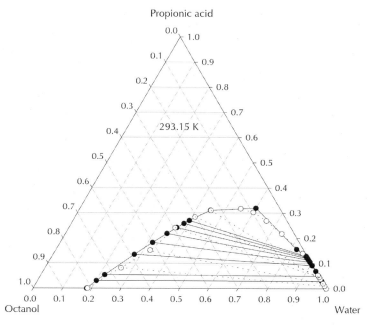

FIGURE 14.4 Correlation of the UNIQUAC data for (water + propionic acid + octanol) system at 293.15 K (○) Experimental points; (•) SAPEN and UNIQUAC points.

TABLE 14.2 Structural parameters r, q.

Composition	r	q
water	0.920	1.40
Proiponic acid	2.8768	2.6120
1-octanol	6.120	5.02

TABLE 14.3 UNIQUAC model prameters

	water	Propionic acid	1-octanol
Water	---------	−140.02	295.79
Propionic acid	295.53	----------	256.82
1-octanol	193.59	−172.02	----------

The root mean-square deviation (RMSD) was calculated from the difference between the experimental and calculated mole fractions according to the following equation:

$$RMSD\% = 100\sqrt{\frac{\sum_{k=1}^{1}\sum_{j=1}^{2}\sum_{r=1}^{3}\left(x_{jk} - x_{jk}\right)^2}{6n}}$$

where n is the number of tie-lines, x indicates the experimental mole fraction, x is the calculated mole fraction, and the subscript I indexes components, j indexes phases and $k = 1,2,\ldots,n$ (lie-lines). The UNIQUAC model was used to predict the experimental data at different temperature with RMSD% value of 12.94 % as reported in Table 14.1.

The ability of 1-octanol to extract propionic acid from water can be determined using the following equation:

$$S = D_a/D_w$$

where D_a is the {mole fraction of acid in organic phase/mole fraction of acid in aqucouse phase) and D_w is the (mole fraction of water in organic phase/mole fraction of water in acueouse phase).

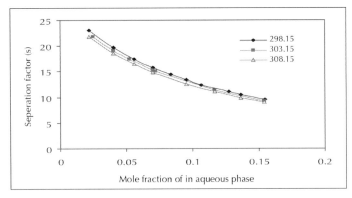

FIGURE 14.5 Seperation factor S of propionic acid as a function of the mole fraction of propionic acid in aqueouse phase at different temperatures.

The consistency of experimental tie-line data can bedetermined using the Othmer and Tobias correlation for the ternary system:

$$\ln[1-x_{33}/x_{33}] = a + b \ln[1 - x_{11}/x_{11}]$$

where a and b are constant, X_{33} is mass fraction of 1-octanol in the extracted phase (organic-rich phase), and X_{11} is the mass fraction of water in aqueous phase. Othmer-Tobias plots were presented in Figure 14.6 for the system at several temperatures and the correlation parameters are listed in Table 14.4. As it can be seen, the plots are linear at each temperature (the correlation factor is close to I ($R^2 = 1$)) indicating a high degree of consistency of the related data.

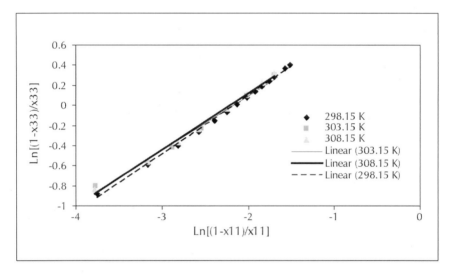

FIGURE 14.6 Othmer—Tobias of the (water + propionic acid + 1-octanol) ternary system at different temperature.

14.4 SIMULATION AND SEPARATION PROGRAM

A commercial simulation program, ASPEN, was used for simulation of the fractional distillation column. The flow diagram for propionic acid extraction process is show in Figure 14.7. The propionic acid concentration used in our design are 20 and 80 wt.%. By removing water from the product fliw, the propionic acid concentration the top of the distillation column is 39.08 wt%. The distillation column was optimized at 12 plates with feed entering at plate 6.

Details of process for the propionic acid purification set together with operation conditions (T,P) are presented in Table 14.5. The operation conditions were selected using the LLE data. It should be noted that, in this table concentration is in terms of weight fraction.

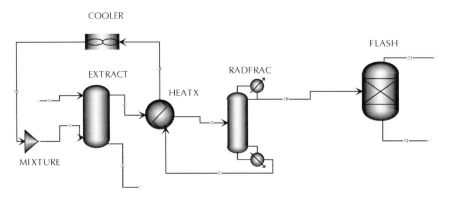

FIGURE 14.7 Propionic acid process separation flow diagram using extraction with 1-octanol.

TABLE 14.4 Othmer—Tobias equation constants for (water + propionic acid + 1- octanol).

Temperature (K)	Othmer–Tobias correlation		
	A	B	R^2
298.15	0.5822	1.2583	1
303.15	0.5413	1.1739	1
308.15	0.5606	1.2402	1

TABLE 14.5 Detailed materials and stream compositions.

Stream ID	1	2	3	4	5	6	7	8	9	10	11	12
Water	0.8	0	0	0.1372	0.1372	1	0	0	0	0.3908	0	1
Propionic acid	0.2	0	0	0.2308	0.2308	0	0	0	0	0.6092	1	0
1-Octanol	0	1	1	0.6320	0.6320	0	1	1	1	0	0	0
Totals (lb/hr)	1102.3	22.046	598.44	946.67	946.67	754.09	576.39	576.39	576.39	370.27	218.47	151.80
Temperature (°F)	77	77	77	76.89	198.04	75.93	562.38	78.69	77	341.49	341.49	341.49
P (psi)	29	58	58	2.9	2.9	2.9	116.03	116.03	58	116.03	116.03	116.03

14.5 CONCLUSION

Tie-line data of the ternary system containing of (water + propionic acid + 1-octanol) were obtained at temperature from (293.15 to 308.15) K. Experimental LLE data of this work analyzed and predicted using UNIQUAC and ASPEN model. The average RMSD value between the observed and calculated mole fractions was 12.94% for the UNIQUAC and ASPEN model. It can be concluded that 1-octanol has high separation factor, very low solubility in water. low cost, high boiling point which may be an adequate solvent to extract propionic acid from its dilute aqueouse solutions.

KEYWORDS

- **Distillation**
- **Fermentation**
- **Herbicides**
- **Propionic acid**

REFERENCES

Abrams, D. S. and Prausnits, J. M. (1975). *AICHE J.* **21**, 116.

Aljimaz, A. S., Fandary, M. S. H., Alkandary, C. A., and Fahim, M. A. (2000). *J. Chem. Eng. Data* **45**(2), 301.

Al-Muhtaseb, S. A. and Fahim, M. A. (1996). *Fluid Phase Equilibr.* **123**, 189–203.

Alkandary, J. A., Aljimaz, A. S., Fandary, M. S., and Fahim, M. A. (2001). *Fluid Phase Equilibr.* **187–188**, 131.

Babuchowski, A., Hammond, E. G., and Glatz, B. A. (1993). Survey of propionibacteria for ability to produce propionic and acetic acids. *J. Food Prot.* **56**, 493–496.

Bendova, M., Rehak, K., Sewry, J. D., and Radioff. S. E. (1994). *J. Chem. Eng. Data* **39**, 320–323.

Carrondo, M. J. T., Crespo, J. P. S. G, and Moura, M. J. (1988). Production of propionic acid using a xylose utilizing Propionibacterium. *Appl. Biochem. Biotechnol.* **17**, 295–312.

Emde R. and Schink, B. (1990). Enhanced propionate formation by Propionibacterium freudenreichii in a three-electrode ampero-metric culture system. *Appl. Environ. Microbiol.* **56**, 2771–2776.

Fahim, M. A., Al-Muhtaseb, S. A., Al-Nashef, I. M. (1997). *J. Chem. Eng. Data* **42**(1), 183.

Fandary, M. S. H., Aljimaz, A. S., Al-Kandary, J. A., and Fahim, M. A. (1999). *J. Chem. Eng. Data* **44**, 1129–1131.

Ghanadzadeh, H. and Ghanadzadeh, A. (2003). *J. Chem. Thermodyn.* **35**, 1393–1401.

Kahya, E., Bayraktar, E., and Mehmeto, G. U. (2001). Optimization of process parameters for reactive lactic acid extraction. *Turk. J. Chem.* **25**, 223–230.

Lewis, P. V. and Yang, S. T. (1992). A novel extractive fermentation process for propionic acid production from whey lactose. *Bio-technol. Prog.* **8**, 104–110.

Magnussen, T. Rasmussen, P., and Fredenslund, A. (1981). *Ind. Eng. Chem. Procces. Des. Dev.* **20**, 331–339.

Martin S. Ray. (1998). Chemical Engineering Design project: A case study Approach, 2nd ed.

Othmer, D. F. and Tobias, P. E. (1942). *Ind. Eng. Chem.* **34**, 693–700.

Pesche, N. And Sandler, S. I. (1995). *J. Chem. Eng. Data* **40**, 315–320.

Playne, M. J. (1985). Propionic and butyric acids. In *Comprehensive Biotechnology*. M. Moo-Young (Ed.). Pergamon, New York, vol. 3, pp 731.

Quesada-Chanto, A., Afschar, A. S., and Wagner, F. (1994). Microbial production of propionic acid and vitamin B12 using molasses or sugar. *Appl. Microbiol. Biotechnol.* **41**, 378–383.

Sandler, S. I. (1994). *Model for thermodynamic and Phase Equilibria calculation*. Dekker, New York.

Sola, C., Casas, C., Godia, F., Poch, M., and Serra, A. (1986). *Biotechnol. Bioeng. Symp.* **17**, 519–523.

Wasewar, K. L., Heesink, A. B. M., et al. (2003), Intensification of Enzymatic Conversion of Glucose to Lactic Acid by Reactive Extraction. *Chem. Eng. Sci.* **58**(15), 3385–3394.

Yang, S., White, S. A., and Hsu, S. (1991). Extraction of carboxylic acids with Tertiary and Quaternary Amines: Effect of pH. *Ind. Eng. Chem. Res.* **30,** 1335–1342.

Zhang, S., Hiaki, T., Hongo, M., and Kojima, K. (1998). *Fluid Phase Equilib. 144*, 97–112.

15 Practical Hints on Optimization of UNIQUAC Interaction Parameters

CONTENTS

15.1 Introduction 147
15.2 Experimental..148
    15.2.1 Materials 148
    15.2.2 Apparatus and Procedure 148
15.3 The Uniquac Model...149
15.4 Results 150
15.5 Conclusion...154
Keywords 154
References...154

15.1 INTRODUCTION

Liquid–liquid equilibrium (LEE) data of ternary systems are required for the design of liquid extraction processes. Also, there is a constant need for phase equilibrium data of these systems for simulation and optimize of separation equipment, valuable information about the molecular interactions, macroscopic behavior of fluid mixtures, and can be used to test and improve thermodynamic models for calculating and predicting fluid-phase equilibria.

A large amount of investigation has been carried out in recent years on the LLE measurements of ternary systems, in order to understand and provide further information about the phase behavior of such systems (Arce et al., 1995; Briones et al., 1994; Dramur and Tatli, 1993; Garcia-Flores et al., 2001; Ghanadzadeh and Ghanadzadeh, 2002, 2003).

In this work, the LLE data for the ternary system of (water + 1-hexanol + TBA) at temperatures from (298.15 to 305.15 K) are presented. Here, TBA is used as a solvent in the separation of 1-hexanol from water. Complete phase diagrams are obtained by solubility and tie-line data simultaneously for each temperature. Selectivity values (S) are also determined from the tie-line data to establish the feasibility of the use of these liquid for the separation of (water + 1-hexanol) binary mixture. The experimental LLE data are correlated using the universal quasi-chemical (UNIQUAC).

15.2 EXPERIMENTAL

15.2.1 Materials

All chemicals used in this work, 1-hexanol and TBA, obtained by Merck with purities >99% and were used without further purification. The purity of these materials checked by gas chromatography. Deionised water was further distilled before use.

15.2.2 Apparatus and Procedure

The binoda (solubility) curves were determined by the cloud point method in an equilibrium glass cell (similar to that of Peschke and Sendler), Figure 15.1, connected to a thermostat was made to measure the LLE data. The temperature of the cell was controlled by a water jacket and maintained with an accuracy of within ± 0.1 K. The mixture was vigorously agitated by a magnetic stirrer for 4 hr. The prepared mixtures were then left to settle for 4 hr for phase separation. The samples of organic-rich phase were taken by a syringe (1 µl) from the upper layer and that of water-rich phase from a sampling tap at the bottom of the cell.

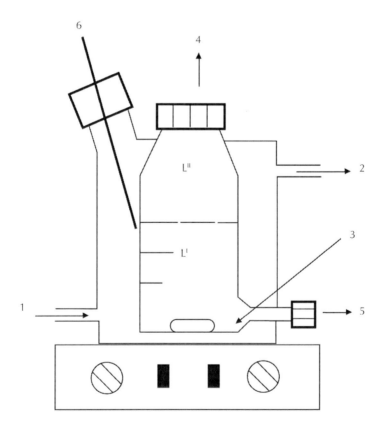

FIGURE 15.1 Fig cell.

The tie-line data were obtained by preparing ternary mixtures of known overall compositions lying within the two-phase region, and after being allowed to reach equilibrium, samples were carefully taken from each phase and analyzed. Both the phases were analyzed using a Varian CP-3,800 gas chromatography (GC) equipped with a thermal conductivity detector (TCD) and Star integrator. A 4 m x 4 mm column packed with CHROMOSORB T 40–60 Mesh was used to separate the components. The injection and the detector temperatures were 250 K. The carrier gas (helium) flow rate was maintained at 40 ml/min.

15.3 THE UNIQUAC MODEL

The experimental LLE data of a ternary system can be correlated using the universal quasi–chemical model (UNIQUAC) (Escudero and Cabezas, 1996; Escudero et al., 1994; Weast, 1989–1990). The mole fractions x_i^E, x_i^R of LLE phases (extracted phase and raffinate phase) can be determined using the following equation:

$$(\gamma_i x_i)^E = (\gamma_i x_i)^R \tag{1}$$

$$\sum x_i^R = 1 \tag{2}$$

Here γ_i^E and γ_i^R are the corresponding activity coefficients of component i in extracted phase and raffinate phase. The UNIQUAC equation for the liquid-phase activity coefficient is represented as follows:

$$\ln \gamma_i = \ln \gamma_i(\text{combinatorial}) + \ln \gamma_i(\text{residual}) \tag{3}$$

The combinatorial and residual parts of the activity coefficient are due to difference in shape and energy of the molecules, respectively. The combinational and residual parts of the coefficient can be written as follows:

$$\ln \gamma_i^C = \ln\left(\frac{\Phi_i}{x_i}\right) + \frac{z}{2} q_i \ln\left(\frac{\theta_i}{\Phi_i}\right) + \iota_i - \frac{\Phi_i}{x_i} \sum_{j=1}^{C} x_j \iota_j \tag{4}$$

$$\ln \gamma_i^R = q_i \left[1 - \ln\left(\sum_{j=1}^{c} \theta_j \tau_{ji}\right) - \sum_{j=1}^{c} \left(\frac{\theta_j \tau_{ij}}{\sum_{k=1}^{c} \theta_k \tau_{kj}}\right) \right] \tag{5}$$

Here τ_{ij} is the adjustable parameter in the UNIQUAC equation. The parameter Φ_i (segment fraction), θ_i (area fraction) and τ_{ij} are given by the following equations:

$$\Phi_i = \frac{x_i r_i}{\sum_{i=1}^{c} x_i r_i} \quad \text{and} \quad \theta_i = \frac{x_i r_i}{\sum_{i=1}^{c} x_i q_i} \tag{6}$$

$$\tau_{ij} = \exp\left(-\frac{\ddot{A}u_{ij}}{RT}\right) = \exp\left(-\frac{a_{ij}}{T}\right) \quad (7)$$

The experimental results were compared with those correlated using the UNIQUAC model and the values for the interaction parameters were obtained for this model The UNIQUAC parameters were estimated by Aspen. The UNIQUAC model has been successfully applied for the correlation of several LLE systems. This model depends on optimized interaction parameters between each pair of components in the system, which can be obtained by experiments. As the optimized interaction parameters can be also correlated to temperature, the interaction parameters in the UNIQUAC model were also investigated.

15.4 RESULTS

The LLE measurements for the ternary system were made at atmospheric pressure in the temperature range at (298.2, 303.2, and 305.2 2) K. The experimental and correlated LLE data of water, 1-hexanol and TBA at each temperature are obtained. Experimental tie line data for (water + 1-heaxol + TBA) at each temperature were reported in Table 15.1.

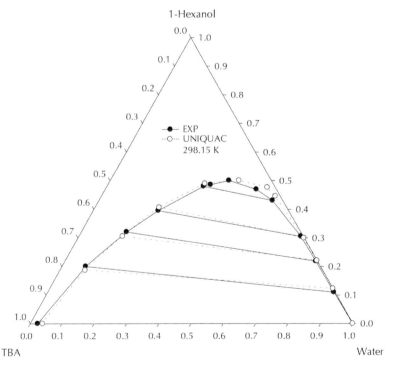

FIGURE 15.2 Experimental (—) and correlated UNIQUAC (- - -) LLE data at 298.15 K.

Practical Hints on Optimization of UNIQUAC Interaction Parameters

TABLE 15.1 Experimental tie line data for (water + 1-hexanol + TBA) at each temperature

Water-rich phase	(mole fraction)		Solvent-rich phase	(mole fraction)	
Water	1-hexanol	TBA	Water	1-hexanol	TBA
T=298.2K					
0.9999	0.0000	0.0001	0.0245	0.0000	0.9755
0.8867	0.1098	0.0035	0.0750	0.2001	0.7249
0.7774	0.2178	0.0048	0.1407	0.3210	0.5385
0.6880	0.3048	0.0072	0.2026	0.3950	0.4014
0.5389	0.4300	0.0311	0.3005	0.4798	0.2197
0.4691	0.4691	0.0618	0.3194	0.4855	0.1951
T=303.2K					
0.9860	0.0000	0.0140	0.0204	0.0000	0.9796
0.9380	0.0611	0.0009	0.0973	0.1639	0.7388
0.7573	0.2409	0.0018	0.2104	0.3354	0.4542
0.6942	0.3009	0.0049	0.2541	0.3766	0.3693
0.6292	0.3594	0.0114	0.3187	0.4130	0.2683
0.5414	0.4169	0.0417	0.3590	0.4273	0.2137
T=305.2K					
0.9948	0.0000	0.0052	0.0303	0.0000	0.9697
0.8816	0.1125	0.0059	0.0934	0.1744	0.7322
0.7716	0.2213	0.0071	0.1427	0.2768	0.5805
0.6897	0.3019	0.0084	0.2090	0.3540	0.0437
0.6145	0.3736	0.0119	0.2936	0.4106	0.2958
0.4959	0.4474	0.0567	0.3659	0.4464	0.1877

The experimental and correlated tie lines for this system at 298.2 K were plotted in Figure 15.2. From the LLE phase diagram, (TBA + water) is the only pair that is partially miscible and two liquid pairs (water + 1-hexanol) and (1-hexanol + TBA) are completely miscible. As it can be seen from Figure 15.2, the phase diagram shows plait point. At this point, only one liquid-phase exists and the compositions of the two phases are equal.

To show the selectivity and strength of the solvent in extracting the acid, distribution coefficients (D_i) for the 1-hexanol (i = 2) and water (i = 1) and the separation factor (S) is determined as follows:

$$D_i = \frac{\text{Weight fraction in solvent phase } (W_{i3})}{\text{Weight fraction in solvent phase } (W_{i1})} \quad (9)$$

$$S = \frac{(D_2)}{(D_1)} \quad (10)$$

The distribution coefficients and separation factor for each temperature are given in Table 15.2. The extraction power of the solvent at each temperature, plot of S vs. X_{21}, is given in Figure 15.3.

TABLE 15.2 Distribution coefficients (Di) of water (1) and 1-hexanol (2) and separation factors at 298.15, 303.15 and 305.15 K..

298.15 K			303.15 K			305.15 K		
D_2	D_1	S	D_2	D_1	S	D_2	D_1	S
1.8224	0.0846	21.5457	2.6825	0.1037	25.8600	1.5502	0.1059	14.6325
1.4738	0.1810	8.1432	1.3923	0.2778	5.0113	1.2508	0.1849	6.7632
1.2959	0.2945	4.4008	1.2516	0.3660	3.4193	1.1726	0.3030	3.8695
1.1158	0.5576	2.001	1.1491	0.5065	2.2687	1.0990	0.4778	2.3003
1.0350	0.6809	1.5200	1.0249	0.6631	1.5457	0.9978	0.7379	1.3523

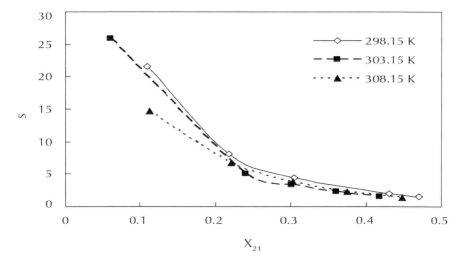

FIGURE 15.3 Separation factor (S) of 1-hexanol as a function of the mole fraction (X21) of 1-hexanol in aqueous phase.

The consistency of experimentally measured tie line data can be determined using the Othmer and Tobias correlation (Othmer and Tobias, 1942) for the ternary system at each temperature:

$$\ln\left(\frac{1-X_{33}}{X_{33}}\right) = A + B\ln\left(\frac{1-X_{11}}{X_{11}}\right) \tag{11}$$

where A and B are constant, X_{33} is mole fraction of TBA in the extracted phase (organic-rich phase) and X_{11} is the mole fraction of water in the raffinate phase (aqueous-rich

Practical Hints on Optimization of UNIQUAC Interaction Parameters 153

phase). Othmer–Tobias plots were presented in Figure 15.4 for the system at several temperatures and the correlation parameters ($R^2 \approx 1$ are listed in Table 15.3. The linearity of the plots indicates the degree of reliability of the related data.

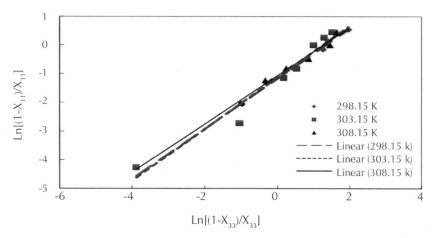

FIGURE 15.4 Othmer–Tobias of the (water + 1-hexanol + TBA) ternary system at different temperatures.

TABLE 15.3 Othmer–Tobias equation constants for (water + 1-hexanol + TBA) ternary system.

	Othmer -Tobias correlation		
T (K)	A	B	R^2
298.2	0.8720	-1.1693	0.9962
303.2	0.9005	-1.1252	0.9579
305.2	0.8396	1.0761	0.9786

The root mean square deviation (RMSD) can be taken as a measure of precision of the correlations. The RMSD was calculated from the difference between the experimental and calculated mole fractions according to the following equation:

$$\text{RMSD\%} = 100\sqrt{\frac{\sum_{k=1}^{n}\sum_{j=1}^{2}\sum_{i=1}^{3}\left(\hat{X}_{ijk}-X_{ijk}\right)^2}{6n}} \tag{12}$$

$\hat{X}$ where n is the number of tie-lines, x and indicate the experimental and calculated mole fraction, respectively. The subscript i indexes components, j indexes phases and k = 1, 2,..., n (tie-lines). The UNIQUAC model was used to correlate the experimental data at each temperature (298.15, 303.15, and 305.15 K) with RMSD% values of 1.42, 1.97, and 1.33%, respectively.

15.5 CONCLUSION

The LLE data of the ternary system composed of water + 1-hexanol + TBA were measured at different temperatures of (298.2, 303.2, and 305.2) K. The UNIQUAC model was used to correlate the experimental LLE data. The optimum UNIQUAC interaction parameters between water, 1-hexanol, and TBA were determined using the experimental liquid–liquid data. The average RMSD value between the observed and calculated mole fractions with a reasonable error was 1.57% for the UNIQUAC model. The solubility of water in TBA increases with amounts of 1-hexanol added to water + TBA mixture.

KEYWORDS

- Liquid–liquid equilibrium
- Root mean square deviation
- Ternary system
- Thermodynamic models
- Universal quasi-chemical

REFERENCES

Arce, A., Blanco, A., Martinez-Ageitos, J., and Vidal, I. (1995). *Fluid Phase Equilib.* **109**, 291–297.

Briones, J. A., Mullins, J. C., and Thies, M. C. (1994). *Ind. Eng. Chem. Res.* **33**, 151–156.

Dramur, U. and Tatli, B. J. (1993). *J. Chem. Eng. Data* **38**, 23–25.

Escudero, I. and Cabezas, J. L. (1996). *J. Chem. Eng. Data* **41**, 2–5.

Escudero, I., Cabezas, J. L., and Coca, J. (1994). *J. Chem. Eng. Data* **39**, 834–839.

Garcia-Flores, B. E., Galicia-Aguilar, G., Eustaquio-Rincon, R., and Trejo, A. (2001). *Fluid Phase Equilib.* **185**, 275–293.

Ghanadzadeh, H. and Ghanadzadeh, A. (2002). *Fluid Phase Equilib.* **202** 337–344.

Ghanadzadeh, H. and Ghanadzadeh, A. (2003). *J. Chem. Thermodynamics* **35** 1393–1401.

Othmer, D. F. and Tobias, P. E. (1942). *Ind. Eng. Chem. Res.* **34**, 690–700.

Weast, R. C. (1989–1990). *Handbook of Chemistry and Physics*, 17th ed., CRC Press, Boca Raton, FL.

16 A Mathematical Approach to Control the Water Content of Sour Gas

CONTENTS

16.1 Introduction 155
16.2 Method...156
    16.2.1 Artificial Neural Networks 156
16.3 Results 158
16.4 Discussion..160
16.5 Conclusion 162
Keywords 162
References..162

16.1 INTRODUCTION

Natural gas reservoirs always have water associated with them; gas in the reservoir is saturated by water. When the gas is produced water is produced too from the reservoir directly. Other water produced with the gas is water of condensation formed because of the changes in pressure and temperature during production. In the transmission of natural gas further condensation of water is troublesome (Lukacs, 1962, 1963). It can enlarge pressure drop in the line and frequently goes to corrosion problems. Therefore, water should be removed from the natural gas before it is offered to transmit in the pipeline. For these argue, the water content of sour gas could be important for engineering attention. In a study of the water content of natural gases Lukacs (1962, 1963) measured the water content of pure methane at 160°F and pressures up to 1,500 psia also Gillespie et al. (1980, 1984) predicted the water content of methane in the range of 122–167°F and for pressures from 200 to 2,000 psia. Sharma (1969) proposed a method for calculating the water content of sour gases, originally designed for hand calculations but it was slightly complicated. Bukacek (1990) suggested a relatively simple correlation for the water content of sweet gas, based on using an ideal contribution and a deviation factor. McKetta et al. published a chart for estimating the water content of sweet natural gas. This chart has been modified slightly over the years and has been reproduced in many publications (GPSA Engineering Data Book, 1998). Recently, Ning et al. (2000) proposed a correlation based on the McKetta et al. chart. This correlation reveals how difficult it can be to correlate something that is as seemingly simple as the water content of natural gas. Maddox (1988) developed a method for estimating the water content of sour natural gas. His method assumes that the water content of sour gas is the sum of three terms sweet gas contribution (Methane, CO_2 and H_2S).

Most of the traditional methods work in the limited range of pressure and temperature and they have a good accuracy in this limited range, which is near the ideal equilibrium condition. But in the high pressure and temperature gases have non-linear behavior that these methods cannot predict the gas behavior (Hugan et al., 1985). The artificial neural networks (ANN), as a good non-linear function approximator, can simulate the non-linear functions with high accuracy (Hagan and Menhaj,1994; Haykin, 1999). In this chapter we predicted the water content of the sour natural gas mixtures with the ANN method. The results show the ANN's capability to predict the measured data. We compare our results with the other numerical and analytical methods, for example Wichert and Bukacek Maddox. These comparisons confirm the superiority of the ANN method.

The outline of this chapter is as follow: In method section, details of new method are derived and we explain how it can be applied for water content prediction. Result section presents our numerical results.

16.2 METHOD

16.2.1 Artificial Neural Networks

The ANN is constructed as a massive connection model of simply designed computing unit, called "neuron". Figure 16.1 illustrates a simple model of n-inputs single-output neuron. All the input signals are summed up as z and the amplitude of the output signal is determined by the non-linear activation function $f(z)$. In this work, we employ the modified sigmoid function $f(z)$ given as follow (Hagan and Menhaj, 1994),

$$f(z) = \frac{1-e^{-kz}}{1+e^{-kz}}. \quad (1)$$

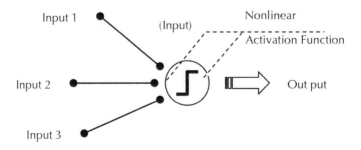

FIGURE 16.1 Basic model of multi-inputs one-output neuron.

Here, we adopt the sigmoid function with moderate slope so that the network can output continuous range of values from −1 to 1, which brings the differentiability of the network (Chong et al., 1999; Hagan and Menhaj, 1994). Here, we adopt a multilayer perceptron (MLP) type network with three layers, which has been used for various applications (Chong et al., 1999; Hagan and Menhaj, 1994; Haykin, 1999). Shown in Figure 16.2 is the architecture of the perceptron neural network. For clear notation, we will use the indices i, j, and k for the units corresponding to "input", "hidden", and "output" layers, respectively (see Figure 16.2). Note also that n_i and o_i are used to represent the input and output to the ith neuron, respectively. Input–output properties of the neurons in each layer can be simply expressed in mathematical term as (Haykin, 1999),

$$o_i = f(n_i), \tag{2}$$

$$o_j = f(n_j), \tag{3}$$

$$o_k = f(n_k), \tag{4}$$

where, as inputs to the neurons are given as,

$$n_i = \text{(input signal to the ANN)}, \tag{5}$$

$$n_j = \sum_{i=1}^{N_i} w_{ij} o_i + \theta_j, \tag{6}$$

$$n_k = \sum_{j=1}^{N_j} w_{jk} o_j + \theta_k. \tag{7}$$

Here, N_i and N_j represent the numbers of the units belonging to "input" and "hidden" layers, while w_{ij} denotes the synaptic weight parameter which connects the neurons i and j.

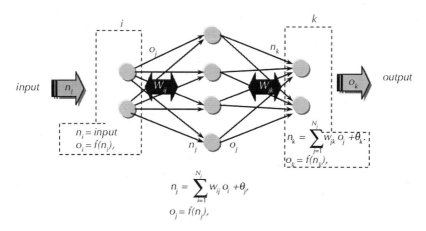

FIGURE 16.2 Multilayer perceptron consisting "input", "hidden", and "output" layers.

The ANN training is an optimization process in which an error function is minimized by adjusting the ANN parameters (weights and biases). When an input training pattern is introduced to the ANN, it calculates an output. Output is compared with the real output (experimental data) provided by the user.

We train the network via the fast convergence gradient-descend back-propagation method with momentum term for the non-negative energy function (Yam and Chow, 2000).

16.3 RESULTS

In this part of our study, the object is to find the optimal performance ANN model for prediction of water content. The results are illustrated in Figure 16.3. In this work, an according Figure 16.3, the optimum number of hidden nodes was selected to be 302. In selecting data for modeling, and to ensure that they represent normal operating ranges, off data were deleted from the data list. The variables of the model and the operating ranges are summarized in Table 16.1.

Data sets were collected from various components in this simulation. For an ANN simulation of gas mixture water content, data sets obtained by Ning et al. (2000), Lukacs (1962, 1963), *GPSA Engineering Data Book* (1998). These data sets are an important sets and suitable for achieving studies on the behavior of natural gases.

Major components used in these references are: methane, propane, hydrogen sulfide, carbon dioxide, and water. In this chapter the analysis data are based on sour gas. Overall 136 data sets were obtained. Among 136 data sets 80 points were used for training the ANN and the remaining 56 data sets were used for accuracy checks of the simulation. The input variables of model and their operating ranges are limited to hydrogen sulfide composition up to 89.6 mol% and is applicable for temperatures between 50 to 350°F and pressure from 200 up to 3,500 psia finally the water content as output of ANN at the range of 40.6 up to 3,500 lb/MMCF.

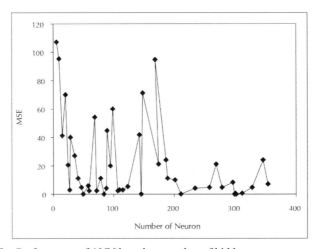

FIGURE 16.3 Performance of ANN based on number of hidden neurons.

A Mathematical Approach to Control the Water Content of Sour Gas

Wichert correlation (Robinson et al., 1978; Wichert and Wichert, 1993) is valid only for H_2S up to 55 mol% in sour gas mixtures. This only covers 68% of data of valid experimental data. In this simulation of gas mixture water content we compared the results of ANN by Wichert and Bukacek Maddox (BM). At this compare BM and Wichert could not predict sour natural gas water content in a wide range of data. These methods have limitations and only can be used in an appointed pressure and temperature. But ANN can predict unknown behaviors of sour natural gases in any limitations of pressure and temperature conditions.

TABLE 16.1 ANN model variables and their ranges.

Variable	range
H2 S (%mol)	7.96–89.52
T (F)	50–350
P (psia)	200–3500
Water Content (lb/MMCF)	40.6–3500

TABLE 16.2 Comparison of the ANN prediction results of the sour natural gas mixture water content with the Wichert and BM methods.

H2S %mol	T(F)	P (psia)	Experimental		
				W.C (calc)	Div%
7.96	200.0	200	2835.0	2506.0	
8.00	130.0	1500	111.0	116.0	
9.06	200.0	200	2820.0	2500.0	
100.0	1100	81.0	75.0	10.00	
15.71	120.0	200	414.8	380.1	
16.00	159.8	1395	226.0	231.0	
17.00	160.0	1010	292.0	294.0	
120.0	200	526.5	379.2	17.46	
18.10	120.0	200	378.8	378.6	
19.00	160.0	611	442.0	418.0	
21.00	16.0	358	712.0	707.0	
27.50	160.0	1392	247.0	264.0	
27.50	160.0	1367	247.0	268.0	
29.00	160.0	925	328.0	330.0	
43.80	120	200	3087.0	2462.0	
75.56	120	200	559.1	.....	
81.25	200	200	2916.0	.....	

TABLE 16.2 (Continued)

H2S %mol	T(F)	P (psia)	Experimental	
			W.C (calc)	Div%
7.96	200.0	200	2835.0	2506.0
89.52	120	200	619.9	.....

AAD%					
Wichert	BM		ANN		
	W.C (calc)	Div%	W.C (calc)	Div %
11.60	2866.0	−1.09	2890.0	−1.94
−4.50	113.0	−1.80	113.0	−1.80
11.34	2855.0	−1.24	2877.9	−2.05
83.0	81.3	−0.37	7.41	−2.47
8.36	430.5	−3.66	420.3	−1.33
−2.21	260.0	−15.04	231.9	−2.61
−0.68	322.0	−10.27	298.6	−2.26
433.7	529.4	−0.55	32.59	17.62
0.05	434.2	−14.62	384.9	−1.66
5.43	467.0	−5.66	447.0	−1.13
0.07	723.0	−1.54	709.3	0.38
253.3	−2.55	−6.88	297.0	−20.24
−8.50	300.0	−21.45	257.2	−4.13
−0.61	375.0	−0.61	334.6	−2.01
31.74	446.0	21.53	570.1	−0.30
3070.1	0.55	20.25	2900.0	6.06
.....	461.7	17.42	561.1	−0.36
.....	2941.0	−0.86	2954.0	−1.30
.....	463.3	25.26	620.0	−0.02

16.4 DISCUSSION

A scatter plot of measured water content against the ANN model predictions is show in Figure 16.4. The prediction, which match measured values, should fall on the diagonal line (line with intercept 0 and slope equal to 1). Almost all data lay on this line, which can confirms the accuracy of the ANN model.

The data points are very close to the diagonal lines and this confirms again the ANN can learn very well relationships between input and output data and generalized successfully of Water Content. Good performance of ANN is obvious when it is compared to other models and simulators. To check the performance of the ANN model,

A Mathematical Approach to Control the Water Content of Sour Gas 161

its estimation and compared with an existing simulator available. The Table 16.2 compares the error of ANN model with Wichert and BM models.

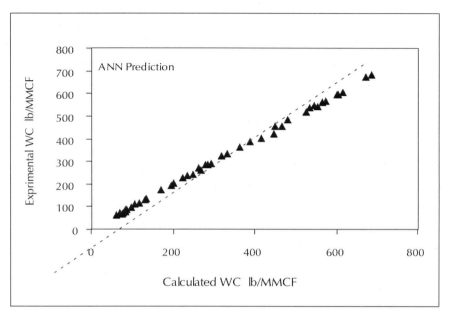

FIGURE 16.4 Artificial Neural Network prediction of gas mixtures water content, (lb/MMCF).

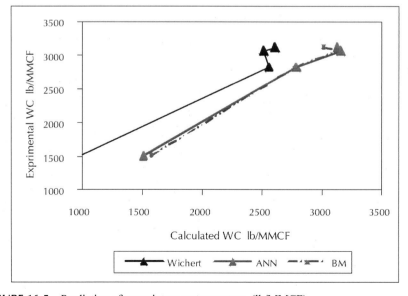

FIGURE 16.5 Prediction of gas mixture water content, (lb/MMCF).

Figure 16.5 compares methods at the range of 1,000–3,500 lb/MMCF of water content. Error of Wichert calculations at this area is more than BM and ANN calculations.

16.5 CONCLUSION

In this chapter ANN model is developed for prediction of natural sour gas mixtures water content. The model is trained based on measured (experimental) data for three various inputs, (H_2S, Pressure, and Temperature).

The difference between ANN model prediction and validation data was very small which confirmed the ability of ANN to accurately predict unseen data. The ANN model was also compared with two numerical and analytical models, Whichert and BM. The results showed that the ANN model accuracy out perform the traditional simulators.

KEYWORDS

- **Artificial neural networks method**
- **Data sets**
- **Natural gas**
- **Pipeline**
- **Sweet gas**
- **Wichert and BM models**

REFERENCES

Bukacek, R. F. and McCain, W. D. (1990). *The Properties of Petroleum Fluids*, 2nd ed., PennWell Books, Tulsa, OK.

Chong, E. K. P. Hui, S., and Stanislaw Zak, H. (1999). An analysis of a class of neural networks for solving linear programming problems. *IEEE transactions on automatic control*, vol. 44, no. 11.

Gillespie, P. C., Owens, J. L., and Wilson, G. M. (1980). Sour Water Equilibria Extended to High Temperature and with Inerts Present AIChE Winter National Meeting, Paper 34-b, Atlanta, GA, Mar. 11-14, (1984) and Gillespie, P.C. and G.M. Wilson, *Vapor–Liquid Equilibrium Data on Water-Substitute Gas Components: N_2-H_2O, H_2-H_2O, CO-H_2O, H_2-CO-H_2O, and H_2S-H_2O* Research Report RR-41, GPA, Tulsa, OK, (1980).

GPSA Engineering Data Book (1998), 11th ed., Gas Processors Suppliers Association, Tulsa, OK.

Hagan, M. T. and Menhaj, M. B. (November, 1994). Training feed-forward neural network with the Marquardt algorithm. *IEEE Transaction on Neural Networks*, vol 5, no.6.

Haykin, S. (1999). *Neural Networks: A Comprehensive Foundation*, 2nd Ed, Prentice-Hall, New York.

Hugan, S. S., Leu, A. D., and Rpbinson, D. B. (1985). The Phase Behavior of Two Mixture of Methane, Carbon, Dioxide, Hydrogen Sulfide and Water. *Fluid Phase Equil.* **19**, 21–23.

Lukacs, J. (1962). Water Content of Hydrocarbon—*Hydrogen Sulphide Gases*, M.Sc. Thesis, Dept. Chem. Eng., University of Alberta, Edmonton, AB.

Lukacs, J. and Robinson, D. B. (1963), Water Content of Sour Hydrocarbon Systems. *Soc. Petrol. Eng. J.* **3**, 293-297.

Luo, Z. (1991). On the convergence of the LMS algorithm with adaptive learning rate for linear feed-forward neural networks, neural computation.

Maddox, R. N. (1974). *Gas and Liquid Sweetening*, 2nd. ed., John M. Campbell Ltd., pp. 39-42.

Maddox, R. N., Lilly, L. L., Moshfeghian, M. and Elizondo, E. (1988). *Estimating Water Content of Sour Natural Gas Mixtures*, Laurance Reid Gas Conditioning Conference, Norman, OK.

Ning, Y., Zhang, H., and Zhou, G. (2000). Mathematical Simulation and Program for Water Content Chart of Natural Gas. *Chem. Eng. Oil Gas* **29**, 75–77 (in Chinese).

Robinson, J. N. Moore, R. A. and Wichart, E. (1978). Chart Help Estimate H_2O Content of Sour Gases. *Oil and Gas J.* 76–78.

Saarineen, S., Bramley R., and Cybenko, G. (1991). The numerical solution of the neural network training problems. CRSD report 1089, Center for Supercomputing Research and Development, University of Illinois, Urbana.

Sharma, S. and Campbell, J. M. (1969). Predict Natural-gas Water Content with Total Gas Usage. *Oil and Gas J.* 136–137.

Wichert, G. C. and Wichert, E. (1993). Chart Estimates Water Content of Sour Natural Gas. *Oil and Gas J.* 61–64.

Yam J. and Chow, T. (2000). A weight initialization method for improving training speed in feedforward neural network. *Neurocomputing* **30**, 219, 232.

17 Modling and Control of Thermodynamic Properties by Artificial Neural Networka (ANNs)

CONTENTS

17.1 Introduction 165
17.2 Artificial Neural Networks (Anns) 166
17.3 Anns Model 167
17.4 Discussion and Result 167
17.5 Conclusion 170
Keywords 171
References 171

17.1 INTRODUCTION

Fluid phase equilibria and mixing properties are of primary interest for theoretical purposes (mathematical model design, parameter estimation, etc.), and for the development of a general proper liquid theory. In chemical industrial processes involving liquid mixtures, the optimization and adequate design of separation

Equipments are conditioned by a sufficient knowledge of mixing thermodynamics (Iglesias et al., 2007). In what is referred to the unit operation field, the optimization of separation operations by extraction or distillation, require knowledge of the two-liquids phase equilibria, and thermodynamics, which can be determined either experimentally or by prediction based on an appropriate model, and a set of data. Artificial neural networks (ANNs) can also predict liquid-liquid equilibrium (LLE) data as well as thermodynamic model and it dose not have the difficulties equation of state, EOS, model for obtain thermodynamic parameter.

Although, EOS are derived based on strong physical principles, there is still certain amount of empiricism involved in terms of several adjustable parameters that are required in mixing rules. Using EOS for estimating the VLE is tedious and requires an iterative method that may sometimes pose problem for real time.

Control of an operating plant. In such cases other faster alternative methods would be more attractive. The development of numerical tools, such as ANN, has paved the way for alternative methods to estimate the LLE (Richardson et al., 2006). Although, efficiency of neural network is already known for application to the chemical engineering processes, only recently ANN have been used in some thermodynamic systems (Laugier and Richon, 2003). The ANNs can extract the desired information directly from experimental data, and need not take into account the detailed information of structures and interactions in the systems, also ANNs can avoid the limitations and improve the prediction accuracy compared to the thermodynamic models and the polynomial correlation method. Ability of ANNs to "learn" and "recognize" highly non-linear and complex relationships find them ideally suited to a wide range of applications in complex systems.

Several authors have reported application of ANN for estimation of thermodynamic properties such as estimation of viscosity, density, vapor pressure, compressibility factor, and phase equlibria. Chouai and Richon (2002) have used a ANN model for estimating the compressibility factor for the liquid and vapor phase as a function of temperature and pressure for several refrigerants. Urata et al. (2002) used ANN to estimate the vapor–liquid equilibrium (VLE) for the binary systems containing hydrofluoroethers (HFEs) and polar compounds. Lagier and Richon (2003) have used ANN model for estimation of compressibility factor and density as a function of pressure and temperature for some refrigerants. Although, a number of papers have been published with experimental data for phase equlibria equilibrium for various systems, not many have used this technique for estimating the VLE (Mohanty, 2006) and not use this thechniqe for LLE system.

17.2 ARTIFICIAL NEURAL NETWORKS (ANNS)

The best example of a neural network is probably the human brain. In fact, the human brain is the most complex and powerful structure known today. The ANNs are composed of simple elements operating in parallel. These elements are inspired by biological nervous systems. The ANN modeling is carried out in two steps; the first step is to train the network whereas the second is to test the network with data, which were not used for training. The unit element of an ANN is the neuron (node). As in nature, the network function is determined largely by the connections between the elements (Richardson et al., 2006).

The ANNs were developed in an attempt to imitate, mathematically, the characteristics of the biological neurons. They are composed by interconnected artificial neurons responsible for the processing of input–output relationships. these relationships are learned by training the ANN with a set of input–output patterns. The ANNs can be used for different purposes; approximation of functions and classification are examples of such applications. The most common types of ANNs used for classification are the feedforward neural networks (FNNs) and the radial basis function (RBF) networks. Probabilistic neural networks (PNNs) are a kind of RBFs that uses a Bayesian decision strategy (Dehghani et al., 2006).

The FNNs are the most frequently used for engineering purposes. They are designed with one input layer, one output layer and hidden layers. The number of neurons in the input and output layers equals the number of inputs and outputs, respectively. The great problem for FNNs is the determination of the ideal number of neurons in the hidden layer(s); few neurons produce a network with low precision and a higher number leads to over fitting. The use of techniques such as Bayesian regularization, together with the Levenberg–Marquardt algorithm, can help overcome this problem.

17.3 ANNs MODEL

The ANNs model is used in this study is the feed forward neural networks. The neural networks for ternary system n-hexane, methanol, and water is based on experimental data reported by Iglesias et al. (2007) in the temperature range 278.15–328.15 K at atmospheric pressure. We have used LLE data at 278.15, 298.15, 308.15, and 328.15 K for training and validating the ANNs model and used LLE data at 288.15 and 318.15 K for testing the model. The best calculated model which optimized in operation has the following specification: it has one hidden layer. The nodes of input are four (three mole fraction with temperature) and the nodes of out put are two (mole fractions). The number of neurons in hidden layer are 20 the tansig—logsig are used as the transfer function in the neurons the momentum coefficient is equal to 0.25.

17.4 DISCUSSION AND RESULT

In the present study, experimental LLE of n-hexane, methanol and water of the ternary system reported by Iglesias et al. (2007) were used to train the model.

In Figure 17.1, the experimental and ANNs data for n-hexane and methanol are compared at a 288.15 K. Figure 17.2 show these data at 318.15 K. These Figures show a good agreement between experimental and ANNs model predition.

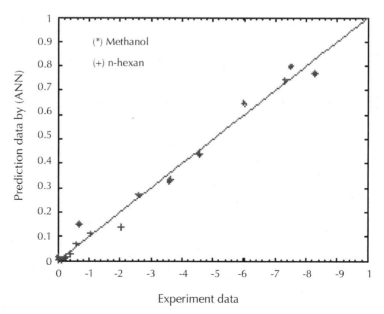

FIGURE 17.1 Comparision of mole fractions experimental and estimated by (ANN) at 288.15K.

168　Chemoinformatics: Advanced Control & Computational Techniques

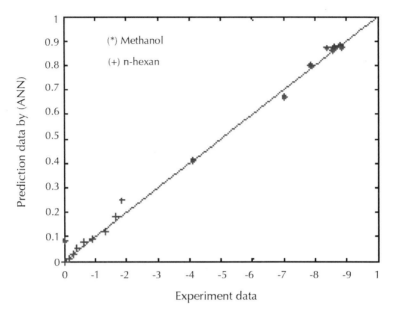

FIGURE 17.2 Comparison of mole fractions experimental and estimated by (ANN) at 318.15K.

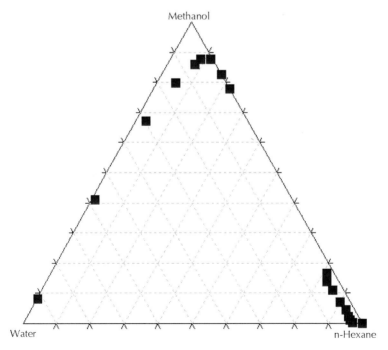

FIGURE 17.3 Prediction equilibria data by ANNs at 288.15 K.

Modling and Control of Thermodynamic Properties

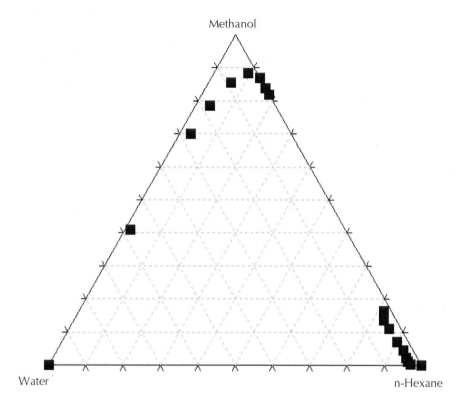

FIGURE 17.4 Experimental equilibria data at 288.15 K.

Figure 3 has been shown triangular diagram of prediction equilibria data by ANNs and Figure 4 is for experimental data of ternary system at 288.15 K.

Table 17.1 and Table 17.2 show the expremental and ANN prediction at 318.15 and 288.15 K respectively for ternary system.

TABLE 17.1 Exprimental and ANN prediction at 318.15 K.

Experimental Data			ANNs prediction		
Y1	Y2	Y3	Y1	Y2	Y3
0	0	1	0.0205	0.0008	0.9803
0.0650	0.0116	0.9234	0.1522	0.0071	0.8549
0.2585	0.0201	0.7214	0.7505	0.0150	0.7445
0.3614	0.0273	0.6113	0.3332	0.0204	0.6872
0.4545	0.0400	0.5055	0.4410	0.0326	0.5915
0.6012	0.0572	0.3413	0.6498	0.0719	0.4224
0.7319	0.1043	0.1638	0.7426	0.1136	0.3710

TABLE 17.1 (Continued)

Experimental Data			ANNs prediction		
Y1	Y2	Y3	Y1	Y2	Y3
0.8255	0.2047	-0.0302	0.7710	0.1391	0.3681
0.7509	0.3627	-0.1136	0.8007	0.3349	0.5343

TABLE 17.2 Expremental and ANN prediction at 288.15 K.

Experimental Data			ANNs prediction		
Y1	Y2	Y3	Y1	Y2	Y3
0	0.0001	0.9999	0.0822	0.0014	0.9192
0.4107	0.015	0.5743	0.4112	0.0111	0.5998
0.7016	0.029	0.2694	0.6716	0.0303	0.3587
0.7864	0.0393	0.1743	0.7989	0.055	0.2562
0.8566	0.06	0.0834	0.8596	0.0794	0.2197
0.8846	0.091	0.0244	0.875	0.0904	0.2154
0.8786	0.1326	-0.0112	0.8767	0.1186	0.242
0.8614	0.1639	-0.0253	0.8734	0.1843	0.3109
0.835	0.186	-0.021	0.8709	0.249	0.3781

We calculate for ANNs model mean absolute error (MAE) and mean square error (MSE). The MSE of ANNs model is compared with MSE of UNIQUIC which is reported by Iglesias et al. (2007). The result shows the ANNs model have a better agreement with the experimental data at higher temperature in see Table 17.3.

TABLE 17.3 MSE and MAE of ANNs and MAE of UNIQUIQ.

Temperature K	328.15	318.15	308.15	298.15	288.15
MSE(UNIQUIC)	0.0533	0.0468	0.0455	0.0255	0.0205
MSE(ANN)	-	0.0445	-	-	0.0306
MAE(ANN)	-	0.0283	-	-	0.0182

17.5 CONCLUSION

In this work, ANNs models have been developed for the LLE ternary system (n-hexane + methanol + water), to estimate the liquid liquid equilibria at the temperature range 278.15–328.15. The weights have been optimized so as to minimize the error between estimated and experimental. The percent deviation in estimating was found to be lower when using ANN model than using UNIQUIC's equation of state at high temperature and in another temperature it can predict equilibria data as well as UNIQUIQ model.

The paper confirms the usefulness of ANNs for numerical description of liquid liquid phase equilibria of the ternary system. Taking into account difficulties in experimental conditions, complicated measurement equipment and unavoidable errors of the devices used which limit the precision of the laboratory measurement results, the accuracy of the results generated by the network may be considered satisfactory for engineering calculations. The properly selected and trained network of relatively simple structure (one hidden layer with 20 neurones) rendered precisely, qualitatively and quantitatively, the thermodynamic character of the ternary system.

It should be remembered that the accuracy of calculation results obtained from an **ANN** is closely related to the accuracy and range of experimental data used for the network learning and testing.

KEYWORDS

- Artificial neural networks
- Equipments
- Hydrofluoroethers
- Mean absolute error
- Mean square error
- Thermodynamic systems
- Vapor–liquid equilibrium

REFERENCES

Chouai, S. L. and Richon, D. (2002). Modeling of thermodynamic properties using neural networks application to refrigerants. *Fluid Phase Equilib.* **199**, 53–62.

Dehghani, M. R., Modarress, H., and Bakhshi, H. (2006). Modeling and prediction of activity coefficient ratio of electrolytes in aqueous electrolyte solution containing amino acids using artificial neural network. *Fluid Phase Equilib.* **244**,153–159.

Iglesias, M., Gonzalez-Olmos, R., Salvatierra, D., and Resa, J. M. (2007). Analysis of methanol extraction from aqueous solution by n-hexane equilibrium diagrams as a function of temperature. *J. Mol. Liq.* **130**, 52.

Laugier, S. and Richon, D. (2003). Use of artificial neural networks for calculating derived thermodynamic quantities from volumetric property data. *Fluid Phase Equilib.* **210**, 247–255.

Mohanty, S. (2006). Estimation of vapor liquid equilibria for the system carbon dioxide–difluoromethane using artificial neural networks. *Int. J. Refrig.* **29**, 243–249.

Richardson, C. J, Mbanefo, A., Aboofazeli, R., Lawrence, M. J., and Brown, D. J. (2006). Prediction of Phase Behavior in Microemulsion Systems Using Artificial Neural Networks. *Coll. Interf. Sci.* **187**, 296–303.

Urata, S., Takada, A., Murata, J., Hiaki, T., and Sekiya, A. (2002). Prediction of vapor–liquid equilibrium for binary systems containing HFEs by using artificial neural network. *Fluid Phase Equilib.* **199**, 63–78.

18 A Study on Liquid–Liquid Equilibria using HYSYS and UNIQUAC Models

CONTENTS

18.1 Introduction 173
18.2 Experimental..174
18.2.1 Materials 174
    18.2.2 Apparatus and Procedure 174
    18.2.3 Uniquac Model 174
18.3 Simulation and Separation Program 181
18.4 Conclusion 182
Keywords 182
References..182

18.1 INTRODUCTION

Phase equilibrium data of ternary systems are very important for simulation, design, and optimization of separation operations. A large amount of investigation has been carried out in recent years on the liquid–liquid equilibria (LLE) measurements of ternary systems, in order to understand and provide further information about the phase behavior of such systems. Since, the liquid extraction of acetic acid from aqueous solution is industrially and scientifically important, the precise LLE data of a liquid mixture composed of (water + acetic acid + organic solvent) are required for further investigations. Various organic solvents for extraction of dilute acetic acid from water have been investigated and reported in the literature (Abrams and Prausnitz, 1975; Aljimaz et al., 2000; Arce et al., 1995, 2001; Bendova et al., 1994; Fandary et al., 1999; Garcia et al., 1988; Garcia-Flores et al., 2001; Ghanadzadeh and Ghanadzadeh, 2003; Ince and Ismail Kirbaslar, 2003; Jassal et al., 1994; Kollerup and Daugulis, 1985; Magnussen et al., 1981; Othmer and Tobias, 1942; Pesche and Sandler, 1995; Sandler, 1994; Sola et al., 1986; Zhang and Hill, 1991; Zhang et al., 1998). Mainly heavy normal alcohols such as hexanol, 1-heptanol, and 1-octanol have been used for extraction of acetic acid from aqueous solution. The LLE data for the mixtures of (water + acetic acid + 2-pentanol), (water + acetic acid + 1-heptanol), and (water + acetic acid + 1-hexanol) had been reported previously by Fahim et al. (Al-Muhtaseb and Fahim, 1996; Aljimaz et al., 2000; Alkandary et al., 2001; Fahim et al., 1997). In the present work 2-ethyl-1-hexanol was used as an organic solvent in order to determined LLE data for the ternary system of (water + acetic acid + 2-ethyl-1-hexanol). Due to its low cost, high boiling point and very low solubility in water, 2-ethyl-1-hexanol has several advantages as a good solvent for recovering of acetic acid form water. Moreover, 2-ethyl-1-hexanol has high separation factor value in the ternary mixture of (water + acetic acid + 2-ethyl-1-hexanol). This solvent also been used as an organic solvent to determine of LLE data for some ternary mixtures of (water +tert-butanol + 2-ethyl-1-hexanol), (water + ethanol+ 2-ethyl-1-hexanol) (Aljimaz et al., 2000; Arce et al., 1995, 2001; Bendova et al., 1994; Fandary et al., 1999; Garcia et al., 1988; Garcia-Flores et al., 2001; Ghanadzadeh and Ghanadzadeh, 2003; Ince and Ismail Kirbaslar, 2003; Jassal et al., 1994; Kollerup and Daugulis, 1985; Pesche and Sandler, 1995; Sola et al., 1986; Zhang and Hill, 1991; Zhang et al., 1998), and (water + acetone + 2-ethyl-1-hexanol) .The aim of this work is to present the phase behavior of LLE for (water + acetic acid + 2-ethyl-1-hexanol). The phase compositions were measured at temperature from (298.2 to 313.2) K. (Martin, 1998)

18.2 EXPERIMENTAL

18.2.1 Materials

The chemicals acetic acid (99.9%) and 2-ethyl-1-hexanol (99.5%) were obtained from Merck and were used without further purification. The purity of these materials was checked by gas chromatography. Distilled water was prepared in our laboratory and used throughout all experiments.

18.2.2. Apparatus and Procedure

The equilibrium data were determined using an experimental apparatus of a 250 ml glass cell, where the temperature of the apparatus controlled by a water jacket and maintained with an uncertainly of within ± 0.001°C. The temperature was measured using a calibrated digital thermometer traceable to the NIST. The mixture was vigorously agitated by a magnetic stirrer for 4 hr. The prepared mixtures were then left to settle for 8 hr for phase separation. The samples of organic-rich phase were taken by a syringe (1 μl) from the upper layer and that of water-rich phase from a sampling tap at the bottom of the cell. Samples were analyzed using Konik gas chromatography (GC), equipped with a thermal conductivity detector (TCD) and Shimadzu C-R2AX integrator. A 2 mm Porapak QS packed column was used to separate the components. The TCD's response was linear and calibrated with 2-ethyl-1-hexanol as in internal standard. The calibration samples were prepared by weighing with an analytical balance accurate to within ± 0.0001 g. The calibration equations were used to convert the area fraction into mole fraction. Calibration coefficients were obtained by fitting a straight line to the calibration results for each composition range. The experimental error of the observed mole fraction during the calibration was about ±0.05%.

18.2.3 UNIQUAC Model

The UNIQUAC model can be used in predicting activity coefficients γ_i. At LLE, the activities of the component i on both phases (extracted phase and raffinate phase) are

A Study on Liquid–Liquid Equilibria using HYSYS and UNIQUAC Models

equal and the mole fractions x_i^E, x_i^R of LLE phases can be determined using the following equations:

$$(\gamma_i X_i)^E = (\gamma_i X_i)^R$$

$$\sum x_i^1 = \sum x_i^2 = 1$$

where γ_i^E and γ_i^R are the corresponding activity coefficients of component i in extracted phase and raffinate phase.

Gibbs free energy can be determined using the following equations:

$$\frac{g^E}{RT} = \sum_{i=1}^{c} x_i \ln(\frac{\phi_i}{x_i}) + \frac{z}{2} \sum_{i=1}^{c} q_i x_i \ln(\frac{\theta_i}{\phi_i}) - \sum_{i=1}^{c} q_i x_i \ln(\sum_{j=1}^{c} \theta_j \tau_{ji})$$

$$\ln \gamma_i = \ln \gamma_i^c + \ln \gamma_i^R$$

$$\ln \gamma_i^c = \ln(\frac{\phi_i}{x_i}) + \frac{z}{2} q_i \ln(\frac{\theta_i}{\phi_i}) + \tau_i - \frac{\phi_i}{x_i} \sum_{j=1}^{c} x_j \tau_j$$

$$\ln \gamma_i^R = q_i \left[1 - \ln(\sum_{j=1}^{c} \theta_j \tau_{ji}) - \sum_{j=1}^{c} (\frac{\theta_j \tau_{ij}}{\sum_{k=1}^{c} \theta_k \tau_{kj}}) \right]$$

where γ_i^c and γ_i^R are the corresponding activity coefficients and miscibility activity coefficients. Z quantity is number and that value is 10.

$$\phi_i = \frac{x_i r_i}{\sum_{i=1}^{c} x_i r_i} \quad \theta_i = \frac{x_i r_i}{\sum_{i=1}^{c} x_i q_i}$$

where Φ_i and Θ_i are the section fraction and area fraction

TABLE 18.1 Experimental and predicted LLE data at each temperature, together with the RMSD% values for [water (1) + acetic acid (2) + 2-ethyl-1-hexanol (3)].

Aqueous phase (raffinate) mole fraction				Organic phase (extract) mole fraction			
x1 (water)		x2 (acetic acid)		x1 (water)		x2 (acetic acid)	
Exp.	UNIQUAC & HYSYS	Exp.	UNIQUAC and HYSYS	Exp.	UNIQUAC and HYSYS	Exp.	UNIQUAC and HYSYS
			298.2K:rmsd%= 8.37				
0.9996	0.9990	0.0000	0.0010	0.1081	0.0355	0.0000	0.0022
0.9739	0.9756	0.0257	0.0244	0.1252	0.0479	0.0936	0.0515
0.9482	0.9489	0.0514	0.0511	0.1331	0.0621	0.1796	0.1022

TABLE 18.1 *(Continued)*

Aqueous phase (raffinate) mole fraction				Organic phase (extract) mole fraction			
x1 (water)		x2 (acetic acid)		x1 (water)		x2 (acetic acid)	
Exp.	UNIQUAC & HYSYS	Exp.	UNIQUAC and HYSYS	Exp.	UNIQUAC and HYSYS	Exp.	UNIQUAC and HYSYS
0.8982	0.8910	0.1016	0.1090	0.1729	0.0947	0.2878	0.2006
0.8821	0.8728	0.1177	0.1271	0.1820	0.1053	0.3113	0.2288
0.8529	0.8430	0.1468	0.1570	0.1960	0.1240	0.3520	0.2740
0.8217	0.8259	0.1776	0.1738	0.2192	0.1346	0.3811	0.2975
0.7957	0.7929	0.2033	0.2056	0.2790	0.1561	0.4230	0.3409
0.6835	0.6840	0.2966	0.3120	0.3970	0.2320	0.4398	0.4520
			303.2K:rmsd%=10.08				
0.9996	0.9990	0.0000	0.0009	0.1081	0.0370	0.0000	0.0023
0.9830	0.9842	0.0168	0.0158	0.1269	0.0453	0.0934	0.0356
0.9483	0.9474	0.0516	0.0526	0.1459	0.0655	0.1868	0.1085
0.8986	0.8940	0.1013	0.1060	0.1763	0.0959	0.2868	0.2005
0.8826	0.8864	0.1172	0.1135	0.1848	0.1004	0.3107	0.2125
0.8533	0.8556	0.1463	0.1443	0.1995	0.1190	0.3476	0.2597
0.8224	0.8205	0.1769	0.1791	0.2116	0.1411	0.3969	0.3091
0.7962	0.7955	0.2027	0.2039	0.2640	0.1574	0.3450	0.3415
0.6960	0.6833	0.2916	0.3121	0.3731	0.2017	0.4788	0.4134
			308.2K:RMSD%= 8.68				
0.9996	0.9990	0.0000	0.0009	0.1081	0.0387	0.0000	0.0024
0.9811	0.9900	0.0186	0.0098	0.1285	0.0438	0.0933	0.0236
0.9484	0.9532	0.0515	0.0468	0.1492	0.0647	0.1864	0.1009
0.8989	0.8930	0.1009	0.1069	0.1797	0.0995	0.2863	0.2066
0.8830	0.8817	0.1167	0.1183	0.1881	0.1062	0.3099	0.2245
0.8538	0.8509	0.1458	0.1490	0.2029	0.1250	0.3468	0.2710
0.8232	0.8227	0.1761	0.1769	0.2360	0.1427	0.4027	0.3102
0.7968	0.7980	0.2001	0.2014	0.2730	0.1588	0.4398	0.3421
0.6750	0.6857	0.3128	0.3098	0.3967	0.2368	0.4679	0.4542
			313.2 K:rmsd%= 7.63				
0.9996	0.9990	0.0000	0.0002	0.1081	0.0399	0.0000	0.0006
0.9793	0.9760	0.0205	0.0241	0.1301	0.0540	0.0929	0.0571
0.9486	0.9302	0.0513	0.0698	0.1280	0.0806	0.1858	0.1476
0.8992	0.8947	0.1006	0.1052	0.1420	0.1014	0.2700	0.2082
0.8835	0.8830	0.1162	0.1169	0.1685	0.1084	0.3320	0.2269
0.8508	0.8536	0.1487	0.1462	0.1820	0.1264	0.3600	0.2713

A Study on Liquid–Liquid Equilibria using HYSYS and UNIQUAC Models

TABLE 18.1 *(Continued)*

Aqueous phase (raffinate) mole fraction				Organic phase (extract) mole fraction			
x1 (water)		x2 (acetic acid)		x1 (water)		x2 (acetic acid)	
Exp.	UNIQUAC & HYSYS	Exp.	UNIQUAC and HYSYS	Exp.	UNIQUAC and HYSYS	Exp.	UNIQUAC and HYSYS
0.8319	0.8360	0.1675	0.1640	0.2022	0.1370	0.3830	0.2960
0.7661	0.7673	0.2243	0.2316	0.2579	0.1821	0.4150	0.3814
0.6815	0.6882	0.2896	0.3074	0.3770	0.2376	0.4190	0.4540

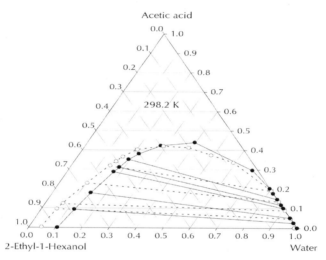

FIGURE 18.1 Prediction of the HYSYS and UNIQUAC data for (water + acetic acid + 2-ethyl-1-hexanol) system at 298.2 K (•) expetimental points; (∘) HYSYS and UNIQUAC points.

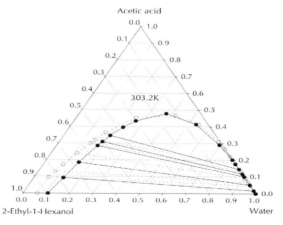

FIGURE 18.2 Prediction of the HYSYS and UNIQUAC data for (water + aceticacid + 2-ethyl-1-hexanol) system at 303.2 K. (•) expetimental points; (∘) HYSYS and UNIQUAC points.

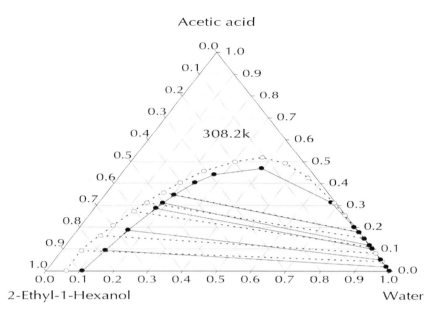

FIGURE 18.3 Prediction of the HYSYS and UNIQUAC data for (water + acetic acid+ 2-ethyl-1-hexanol) system at 308.2 K. (•) expetimental points; (◦) HYSYS and UNIQUAC points.

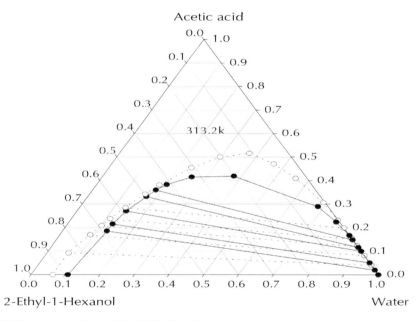

FIGURE 18.4 Prediction of the HYSYS and UNIQUAC data for (water + acetic acid+ 2-ethyl-1-hexanol) system at 308.2 K. (•) expetimental points; (◦) HYSYS and UNIQUAC points.

A Study on Liquid–Liquid Equilibria using HYSYS and UNIQUAC Models 179

The values of the UNIQUAC parameters for LLE calculated by HYSYS program.

TABLE 18.2 Structural parameters r, q.

Composition	r	q
water	0.920	1.40
Acetic acid	2.202	2.072
2-ethyle-1-hexzanol	6.151	5.208

TABLE 18.3 UNIQUAC model parameters.

	water	Acetic acid	2-ethyle-1-hexzanol
Water	------	427.741	585.594
Acetic acid	−305.452	------	46.726
2-ethyle-1-hexzanol	475.22	−34.512	------

The experimental and HYSYS-UNIQUAC LLE data for (water + acetic acid + 2-ethyl-1-hexanol) at different temperature of (298.2 to 313.2) K, are presented in Table 18.1. The estimated uncertainties in the mole fraction were about 0.0005. From the LLE phase diagrams (Figures 18.1–18.4), (2-ethyl-1-hexanol + water) mixture is the only pair that is partially miscible and two liquid pairs (acetic acid + water) and (acetic acid + 2-ethyl-1-hexanol) are completely miscible. The mutual solubility of 2-ethyl-1-hexanol and water is very low and therefore, the high boiling point solvent (2-ethyl-1-hexanol) can be used as a reliable organic solvent for extraction of acetic acid from dilute aqueous solutions.

The root-mean-square deviation (RMSD) was calculated from the difference between the experimental and calculated mole fractions according to the following equation:

$$rmsd\% = 100\sqrt{\frac{\sum_{K=1}^{n}\sum_{j=1}^{2}\sum_{i=1}^{3}\left(\hat{x}_{ijk} - x_{ijk}\right)^2}{6n}}$$

where n is the number of tie-lines, x indicates the experimental mole fraction, $\hat{x}$ is the calculated mole fraction, and the subscript i indexes components, j indexes phases and $k = 1,2,\ldots,n$ (tie-lines). The UNIQUAC model was used to predict the experimental data at different temperature with RMSD% value of 8.3% as reported in Table 18.1.

The ability of 2-ethyl-1-hexanol to extract acetic acid from water can be determined using the following equation:

$$S = \frac{D_a}{D_W}$$

where D_a is the mole fraction of acid in organic phase/mole fraction of acid in aqueous phase and D_w is the mole fraction of water in organic phase/mole fraction of water in aqueouse phase.

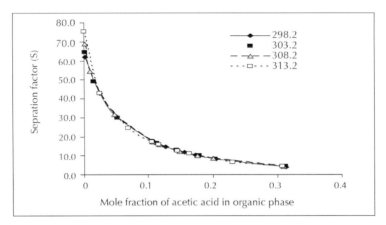

FIGURE 18.5 Separation factor (S) of acetic acid as a function of the mole fraction of acetic acid in organic phase at different temperatures.

The consistency of experimental tie-line data can be determined using the Othmer and Tobias correlation for the ternary system:

$$\ln\left[\frac{1-x_{33}}{x_{33}}\right] = a + b\ln\left[\frac{1-x_{11}}{x_{11}}\right]$$

where a and b are constant, X_{33} is mass fraction of 2-ethyl-1-hexanol in the extracted phase (organic-rich phase) and X_{11} is the mass fraction of water in aqueous phase. Othmer–Tobias plots were presented in Figure 18.6 for the system at several temperatures and the correlation parameters are listed in Table 18.4. As it can be seen, the plots are linear at each temperature (the correlation factor is close to 1 ($R^2 = 1$)) indicating a high degree of consistency of the related data.

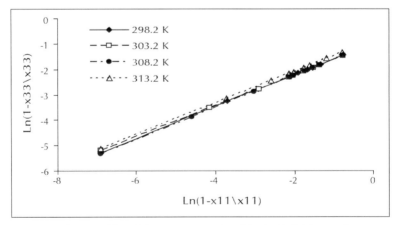

FIGURE 18.6 Othmer–Tobias of the (water + acetic acid + 2-ethyl-1-hexanol) ternary system at different temperatures.

A Study on Liquid–Liquid Equilibria using HYSYS and UNIQUAC Models

TABLE 18.4 Othmer–Tobias equation constants for (water + acetic acid + 2-ethyl-1-hexanol).

Temperature (k)	Othmer–Tobias correlation		
	A	B	R^2
298.2	0.628	-0.9507	1
303.2	0.6088	-1.007	1
308.2	0.6324	-0.9689	1
313.2	0.6229	-0.8412	1

18.3 SIMULATION AND SEPARATION PROGRAM

A commercial simulation program, HYSYS, was used for simulation of the fractional distillation column. The flow diagram for acetic acid extraction process is shown in Figure 18.7. The acetic acid concentration used in our design are 20 and 80 wt%. By removing water from the product flow, the acetic acid concentration the top of the distillation column is 38.6 wt%. The distillation column was optimized at 8 plates with feed entering at plate 4.

Details of process for the acetic acid purification set together with operation conditions (T, P) are presented in Table 18.5. The operation conditions were selected using the LLE data. It should be noted that, in this table concentration is in terms of weight fraction.

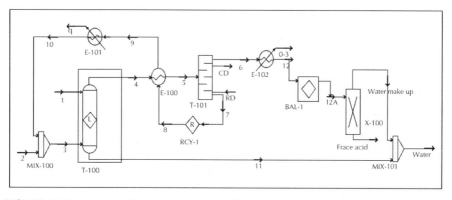

FIGURE 18.7 Acetic acid process separation flow diagram using extraction with 2-ethyl–1-hexzanol.

TABLE 18.5 Detailed materials and stream compositions.

Stream ID	1	2	3	4	5	6	7	8	9	10	11	12	13	14	15
ACETIC ACID	0.2	0.00	0.00	0.2116	0.2116	0.386	0.00	0.00	0.00	0.00	0.00	0.386	0.00	1	0.00
WATER	0.8	0.00	0.00	0.3437	0.3437	0.614	0.00	0.00	0.00	0.00	1.0	0.614	1.0	0.00	1.0
2EH	0.0	1.0	1.0	0.4402	0.4402	0.00	1.0	1.0	1.0	1.0	0.00	0.00	0.00	0.00	0.00

TABLE 18.5 *(Continued)*

Stream ID	1	2	3	4	5	6	7	8	9	10	11	12	13	14	15
Totals(kg/h)	500	10	1004	1340	1340	335.7	1004	994.1	994.1	994.1	164.4	335.7	108.4	227.2	272.8
Temperature (°c)	25	25	25	25	109.5	81.91	201.2	201.2	77.76	25	24.9	25	25	25	25
P(kpa)	101	101	101	100	100	50	200	200	200	200	102	100	100	100	100

18.4 CONCLUSION

Tie-line data of the ternary system containing of (water + acetic acid + 2-ethyl-1-hexanol) were obtained at temperature from (298.2 to 313.2) K. Experimental LLE data of this work were analyzed and predicted using UNQUAC and HYSYS model. The average RMSD value between the observed and calculated mole fractions was 8.3% for the UNIQUAC and HYSYS model. It can be concluded that 2-ethyl-1-hexanol has high separation factor, very low solubility in water, low cost, high boiling point which may be an adequate solvent to extract acetic acid from its dilute aqueous solutions.

KEYWORDS

- **Chromatography**
- **Distilled water**
- **Extracted phase**
- **Raffinate phase**

REFERENCES

Abrams, D. S. and Prausnitz, J. M. (1975). *AICHE J.* **21**, 116.

Aljimaz, A. S., Fandary, M. S. H., Alkandary, C. A., and Fahim, M. A. (2000). *J. Chem. Eng. Data* **45**(2), 301.

Alkandary, J. A., Aljimaz, A. S., Fandary, M. S., and Fahim, M. A. (2001). *Fluid Phase Equilibr.* **187–188**, 131.

Al-Muhtaseb, S. A. and Fahim, M. A. (1996). *Fluid Phase Equilibr.* **123**, 189–203.

Arce, A., Matrinez-Ageitos, J., Rodriguze, O., and Soto, A. (2001). *J. Chem. Thermodyn.* **33**, 139–146.

Arce, A., Matrinez-Ageitos, J., Rodriguze, O., and Vidal, I. (1995). *Fluid Phase Equilib.* **109**, 291–297.

Bendova, M., Rehak, K., Sewry, J. D., Radioff, S. E. (1994). *J. Chem. Eng. Data* **39**, 320–323.

Fahim, M. A., Al-Muhtaseb, S. A., and Al-Nashef, I. M. (1997). *J. Chem. Eng. Data* **42**(1), 183.

Fandary, M. S. H., Aljimaz, A. S., Al-Kandary, J. A., and Fahim, M. A. (1999). *J. Chem. Eng. Data* **44**, 1129–1131.

Garcia, I. G., Perez, A. C., and Calero, F. C. (1988)_. *J. Chem. Eng. Data* **33**, 468–472.

Garcia-Flores, B. E., Galicia-Aguilar, G., Eustaquio-Rincon, R., and Trejo, A. (2001). *Fluid Phase Equilib.* **185**, 275–293.

Ghanadzadeh, H. and Ghanadzadeh, A. (2003). *J. Chem. Thermodyn.* **35**, 1393–1401.

Ince, E., and Ismail Kirbaslar, S. (2003). *J. Chem. Thermodyn.* **35**, 1671–1679.

Jassal, D. S., Zhang, Z., and Hill, G. A. (1994). *Can. J. Chem. Eng.* **72**, 822–826.

Kollerup, F. and Daugulis, A. J. (1985). *Can. J. Chem. Eng.* **63**, 919–927.

Magnussen, T., Rasmussen, P., and Fredenslund, A. (1981). *Ind. Eng. Chem. Process Des. Dev.* **20**, 331–339.

Martin S. Ray. (1998). Chemical Engineering Design Project: A case study Approach, 2nd Edition.

Othmer, D. F. and Tobias, P. E. (1942). *Ind. Eng. Chem.* **34**, 693–700.

Pesche, N. and Sandler, S. I. (1995). *J. Chem. Eng. Data* **40**, 315–320.

Sandler, S. I. (1994). Model for thermodynamic and Phase Equilibria calculation. Dekker, New York.

Sola, C., Casas, C., Godia, F., Poch, M., and Serra, A. (1986). *Biotechnol. Bioeng. Symp.* **17**, 519–523.

Zhang, S., Hiaki, T., Hongo, M., Kojima, K. (1998). *Fluid Phase Equilib.* **144**, 97–112.

Zhang, Z. and Hill, G. A. (1991). *J. Chem. Eng. Data* **36**, 453–456.

19 Optimization and Control of Laboratory Production of Ethanol

CONTENTS

19.1 Introduction 185
19.2 Culture Media and Batch Cultures 186
      19.2.1 Microorganism...186
      19.2.2 Preculture Medium 186
      19.2.3 Fermentation Medium 186
      19.2.4 Airlift Bioreactor..187
      19.2.5 Analytical Methods..188
      19.2.6 Experimental Design 188
19.3 Discussion And Results 189
Keywords 196
References..196

19.1 INTRODUCTION

Cheese whey is a yellowish liquid remaining after milk coagulates during cheese production. It is a by-product of the manufacture of cheese and has several commercial uses.

Cheese whey is produced in huge amounts and is a significant environmental problem due to the high levels of organic matter content (Kosikowski, 1979). Cheese whey represents a biochemical oxygen demand (BOD) of 30–50 g/l and a chemical oxygen demand (COD) of 60–80 g/l lactose is largely responsible for the high BOD and COD, since protein recovery reduces only about 12% of the whey COD (Domingues et al., 1999a; Siso, 1996; Siso et al., 1996).

On the other hand, whey retains much of the milk nutrients, including functional proteins and peptides, lipids, lactose, minerals and vitamins and therefore has a vast potential as a source of added value compounds, challenging the industry to face whey surplus as a resource (OECD-FAO Agricultural Outlook, 2008; Smithers, 2008). In Iran, about 1.8 million tons of whey which is the by-product of cheese producing factories is produced each year. The changing of whey into alcohol, due to the low price of whey (compare to the price of other raw materials), has become the focus of considerable attention in the world. Use of whey in the preparation of ethanol was studied since 1940. Moulin et al. (1980) have achieved to the 86–90% of efficiency in the medium of cheese whey permeate (CCWP) by using the two species of yeas *Kluyveromyces fragilis* and *Candida pseudotropicalis* (Moulin et al., 1980). In a study by Gawel et al. (1978) with *K.fragilis* strain obtained to 10% of ethanol fermentation from whey in 15 days (Gawel and Kosikowski, 1978). Janssens et al. (1984) reported ethanol productivity of 7.1 gl^{-1}h^{-1} for *K. fragilis* operating with cell recycling at $D = 0.15$ hr^{-1} and for CCWP with 100 g/l lactose (Janssens et al., 1984). Terrel et al. (1984) reported ethanol productivity of 13.6 gl^{-1}hr^{-1} for CCWP with 150 g/l lactose concentration operating at continuous operation (Terrel et al., 1984). Ryu et al. (1991) reached to the rate of 2.1 percent of ethanol in a batch system in airlift bioreactor by using *K. fragilis* (20l) (Ryu et al., 1991). Ferrarie et al. (1994) were obtained 64 g/l of ethanol in a fed-batch system (Ferrari et al., 1994). In this research the laboratory production of ethanol by means of whey has been accomplished in an airlift bioreactor. The purposes of the present experimental study were mainly to investigate how the main operating parameters affect the process so as to determine which of them were certainly important. The goals were satisfied by means of response surface methodology through accurately designed central composite design.

19.2 CULTURE MEDIA AND BATCH CULTURES

19.2.1 Microorganism

Yeast strain used in this study was *K. fragilis* PTCC 5193, obtained from the Iranian Research Organization for Science and Technology (IROST). *K. fragilis* was maintained in agar (65gl^{-1}). The culture was sterilized in autoclave at 121°C for 20 min; the yeast inoculum was spread on the surface and incubated at 30°C for 48 hr. At completed growth, the slants were preserved at 4°C.

19.2.2 Preculture Medium

The preculture medium was 13 g/l nutrient broth, 10 g/l peptone, and 10 g/l yeast extract, sterilized at 121°C for 15 min and prepared with a single colony withdrawn from the slants and maintained for 48 hr at an incubator shaker with a temperature of 30°C and velocity of 150 rpm. In all the experiments 100 ml of sterile Erlenmeyer flask were charged with a 50 ml of pre-culture.

19.2.3 Fermentation Medium

Cheese whey (CW) was the fermentation medium; containing lactose (4.5–5% w/v), soluble proteins (0.6–0.8% w/v), lipids (0.4–0.5%w/v), and mineral salts (8–10% of dried extract). Whey also contains appreciable quantities of other components, such as lactic (0.05% w/v) and citric acids, non-protein nitrogen compounds (urea and uric acid) and B group vitamins (Siso, 1996; Siso et al., 1996). For batch experiments 500 ml of erlenmeyer flasks were charged with 300 ml of cheese sterilized and deprotenized whey. It consisted of yeast extract (5 g/l), peptone (5 g/l), NH$_4$Cl (2 g/l), KH$_2$PO$_4$ (1 g/l), MgSO$_4$.7H$_2$O (0.3 g/l) (Domingues et al., 2001; Silveira et al., 2005). Variables were pH, initial lactose concentration (L), yeast cells concentrations (Y), and Na-thioglycolate concentration (NTC) as the reducing agent.

19.2.4 Airlift Bioreactor

Figure 19.1 shows an airlift bioreactor contains external loop which is made of Pyrex glass. The bioreactor was fed with sweet cheese sterilized and deprotenized whey. The cell suspension was aseptically transferred to the bioreactor. Airlift bioreactor was operated at working volume of 7 liters that included 10% pre-culture. The regulation system allows for: temperature control at 30 ± 1°C; foam-level and pH controlled by addition of antifoam and ammonia, respectively. The set-point fixed at pH 5.0 ± 0.1. The system was aerated with filtered air at a different flow rate of 0.1, 0.4, and 0.8 vvm that was controlled using an aeration pump controller. Each run was achieved in duplicates; the average values of lactose, ethanol, and biomass concentrations were calculated and monitored with respect to time.

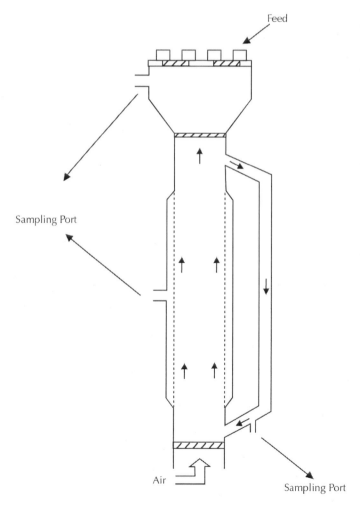

FIGURE 19.1 Schematic of the airlift bioreactor.

19.2.5 Analytical Methods

The samples were removed from the sampling ports at different heights of the column every day and centrifuged at 8,000 rpm for 20 min to remove solids from the liquid media. Analyses were carried out on the supernatants after centrifugation in the Erlenmeyer flask. After 48 hr all of the samples were removed and centrifuged. Total reducing sugar concentrations were measured by using the phenol-acid method (Dubois et al., 1956). The samples were analyzed in triplicates 3%. Ethanol concentrations were measured using a Varian CP-3800 gas chromatograph (GC) equipped with a thermal conductivity detector (TCD), and star integrator. A 2m × 1.4 in × 4 mm column packed with propack Q 89-100 Mesh. The column temperature was set for 75°C for 1 min and raised to 130°C with a rate of 20°C/min yielding a total hold time of 4.75 min. Temperatures of injector and detector were 150 and 200°C, respectively. Dry cell mass concentration was estimated by measuring the optical density of the sample at 600 nm in a spectrophotometer, and by its correlation with the dry cell weight (DCW) obtained gravimetrically. The pH was measured using a pH meter (JEYWAY 3510). The yeast cells concentration was estimated by the dry weight method. The dry weight cell concentration was determined by filtering the sample through 0.2 µm filter paper and then dried at 105°C for 48 hr (Domingues et al., 2001).

19.2.6 Experimental Design

The present experimental study consisted of two steps:

(1) The central composite design (CCD) aimed at determining the effects of four factors on the fermentation process.
(2) Determining the effect of aeration rate on the fermentation process and measuring the variation of ethanol, lactose, and biomass concentration with time in airlift bioreactor.

Four factors were considered to perform for response surface methodology of CCD: pH, initial lactose concentration (L), yeast cells concentration (Y), and Na-thioglycolate concentration (NTC), with three different levels for each of the factors. Values of the factors set equal to 3.5, 5, and 6.5 for pH, 20, 40, and 60 g/l for initial lactose concentration, 0, 100, and 200 mg/l for Na-thioglycolate and 0.03, 0.3, and 0.6 g/l for yeast cells concentration. The ranges of these values were considered since it characterized the optimum range for the yeast activity and the expected range in which the process could be operated. In this study, the experimental design consisted of 15 runs and the independent variables were studied at three different levels. Table 19.1 shows the experimental design used for this study. All the experiments were done in duplicates and the average of ethanol production obtained was taken as the response function (RF). The Second degree polynomials, Eq.(1), which contains all interaction terms, were used to calculate the predicted response:

$$RF = \beta_0 + \Sigma \beta_i x_i + \Sigma \beta_{ii} x_i^2 + \Sigma \beta_{ij} x_i x_j \tag{1}$$

where Y represents response variable, β_0 is the interruption coefficient, β_i the coefficient of the linear effect, β_{ii} the coefficient of quadratic effect and β_{ij} the ijth coefficient of interaction effect, $x_i x_j$ are input variables which influence the response variable Y; β_i is the ith linear coefficient. Numerical analysis of the model was performed to evaluate

19.3 DISCUSSION AND RESULTS

Table 19.1 shows the experimental design and results of CCD of response surface methodology. The factors levels are 3.5, 5, and 6.5 for pH, 20, 40, and 60 g/l for L, 0, 100, and 200 mg/l for NTC, and 0.03, 0.3, and 0.6 g/l for Y. In the last column, the obtained RF values are shown. The effects of the parameters on the RF were calculated and the parameters which showed P-values less than 0.05 were taken into account in the model; the other parameters were actually undistinguishable from noise. The significant terms, as shown in Table 19.2, are pH, Y, NTC, pH², and pH-L. Then, by eliminating the other terms from the model (except L to support hierarchy as requested from the methodology (Box et al., 1978)), Eq. (2) was obtained as a function of the significant factors.

$$RF=1.3725+0.1687PH+0.0887L+0.1887NTC+0.2937Y-0.2825PH^2-0.300PH\text{-}L \quad (2)$$

TABLE 19.1 Experimental design and CCD results of response surface methodology.

Run	A	B	C	D	PH	Lactose (g/l)	Na-thioglycolate (mg/l)	Yeast cells concentration(Y) (g/l)	Response Function (RF)
1	-1	-1	0	0	3.5	20	100	0.3	0.4
2	1	1	0	0	6.5	60	100	0.3	1
3	1	-1	0	0	6.5	20	100	0.3	1.5
4	-1	1	0	0	3.5	60	100	0.3	1.1
5	0	0	1	-1	5	40	0	0.03	0.9
6	0	0	1	-1	5	40	200	0.03	1.3
7	0	0	1	1	5	40	0	0.6	1.6
8	0	0	1	1	5	40	200	0.6	1.9
9	-1	0	0	-1	3.5	40	100	0.03	0.75
10	1	0	0	-1	6.5	40	100	0.03	0.8
11	-1	0	0	1	3.5	40	100	0.6	1.2
12	1	0	0	1	6.5	40	100	0.6	1.5
13	0	-1	1	0	5	20	0	0.3	0.9
14	0	1	1	0	5	60	0	0.3	1.2
15	0	-1	1	0	5	20	200	0.3	1.3

TABLE 19.2 Coefficient estimates in second-order model.

Model term	Coefficient	Computed t-value	p-value
Constant	1.3725	21.942	0.000
PH	0.1687	3.413	0.027
L	0.0887	1.517	0.204
NTC	0.1887	3.226	0.032
Y	0.2937	5.940	0.004
PH²	-0.2825	-3.688	0.021
L²	-0.1175	-1.543	0.200
NTC²	-0.0326	-0.492	0.232
Y²	0.0456	0.0683	0.356
PH-L	-0.3000	-4.290	0.013
PH-NTC	0.0000	0.000	1.000
PH-Y	0.0625	0.894	0.422
L-NTC	-0.0225	-0.240	0.822
L-Y	0.0000	0.000	1.000
NTC-Y	0.0000	0.000	1.000

R^2=96.04% and adjusted R^2=86.13%

Figure 19.1a presenting the response surface performance as a function of both pH and L. The best conditions were achieved at pH 5 and L 40 g/l. With increasing lactose level of 40 to 60 g/l, RF decreased because of cellular osmotic pressure limiting. Figure 19.1b reports the response surface versus pH and NTC. It is evident that a pH value around 5 and NTC around 200 mg/l improves the fermentation process. Optimal extent of NTC as a reducing agent is 200 mg/l. It prevents absorption of oxygen and anaerobic fermentation conditions to provide. Figure 19.2a shows the effects of pH and Y factors on the RF; the best conditions were obtained at pH 5 and Y 0.6 g/l. Figure 19.2b showing the RF dependence on both L and Y; a relatively weak effect of L and a stronger effect of Y can be noted. Figure 19.3a presents the response surface versus L and NTC. The best conditions were achieved at L 40 g/l and NTC 200 mg/l. Figure 19.3b exhibits a strong response surface dependence on both Y and NTC. Moreover, a good system behavior corresponding to a RF of 1.8 is obtained at Y 0.6 g/l and NTC 200 mg/l. The RSM showed that the best set of operating conditions as following: pH 5, L 40 g/l, Y 0.6 g/l, and NTC 200 mg/l. Figures 19.2 and 19.3 show the effect of aeration rate on the ethanol production and the time course of cell, lactose, and ethanol concentrations, using ordinary whey media as substrate, in erlenmeyer and airlift bioreactor cultures.

Optimization and Control of Laboratory Production of Ethanol

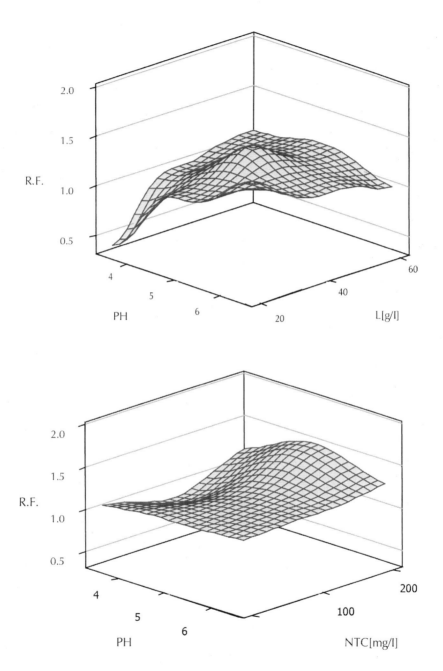

FIGURE 19.2 (a) Response surface as a function of pH and lactose (L). (b) Response surface as a function of pH and Na-thioglycolate (NTC), (the other variables were fixed at middle level).

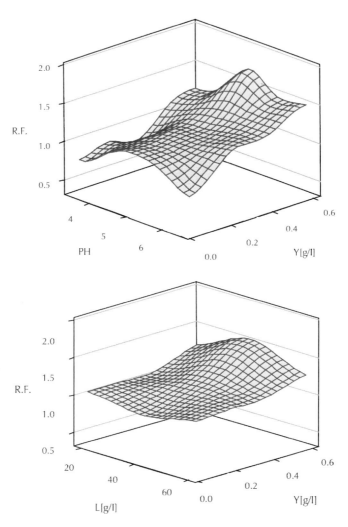

FIGURE 19.3 (a) Response surface as a function of pH and yeast cells concentration (Y). (b) Response surface as a function of lactose (L) and yeast cells concentration (Y), (the other variables were fixed at middle level).

Figure 19.2a shows the maximum amount of alcohol produced in non-aerated conditions in Erlenmeyer, after 65 hr from when fermentation started, is 2.9 (wv^{-1}) percent. Figure 19.2b shows that ethanol production is 2.9 (wv^{-1}) percent after 36 hr for aeration rate of 0.1 vvm in airlift bioreactor. Comparing these two figures shows that amount of ethanol productions are as the same clearly but the duration time of Figure 19.2b to reach the maximum amount of alcohol is shorter, because the aeration operation makes the yeast growth faster. Figure 19.3a shows the ordinary whey media fermentation with 0.4 vvm aeration rate in the bioreactor. The maximum amount of alcohol production after 17 hr was obtained 3 (w/v) percent. Figure 19.3b shows that the highest value of alcohol production is 2.2 (w/v) percent with 0.8 vvm aeration rate in the 19th hr of the fermentation time. As can be seen with the increase of aeration rate of 0.1–0.4 vvm, the amount of alcohol remained approximately constant, but the duration time to reach the maximum amount of alcohol was lower and with further increase of aeration rate of 0.4–0.8 vvm, the alcohol production rate was decreased. Because the fermentation process was anaerobic in nature but the yeast to grow needs a small amounts of oxygen and the excess of the required was reduced the rate of production. The optimum aeration rate for alcohol production is 0.4 vvm. Figure 19.4 illustrates that the amount of alcohol produced is 6.1 (w/v) percent in concentrated whey media of 100 g/l lactose and 0.4 vvm aeration rate. The use of yeast extract (5 g/l), peptone (5 g/l), NH_4Cl (2 g/l), KH_2PO_4 (1 g/l), $MgSO_4 \cdot 7H_2O$ (0.3 g/l), and lactose (50 g/l) increased amount of alcohol considerably (Approximately 3%).

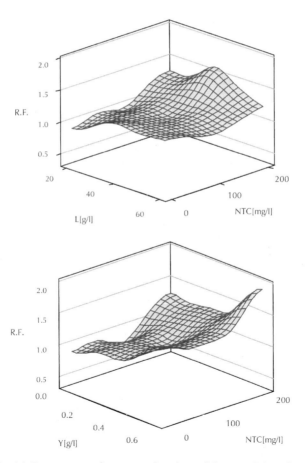

FIGURE 19.4 (a) Response surface as a function of lactose (L) and Na-thioglycolate concentration (NTC). (b) Response surface as a function of yeast cells concentration(Y) and Na-tioglycolate concentration (NTC), (the other variables were fixed at middle level).

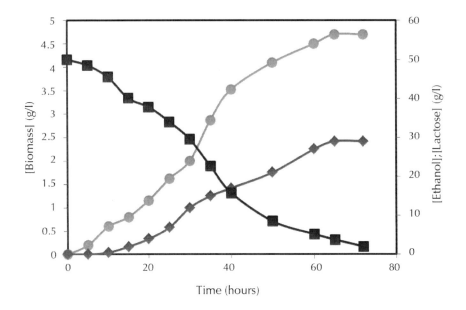

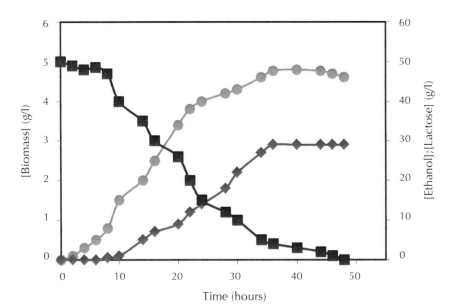

FIGURE 19.5 Cell (●), lactose (■) and ethanol (♦) concentration profiles for ordinary whey media with 50 g/l lactose without (a) with (b) aeration rate of 0.1 vvm.

Optimization and Control of Laboratory Production of Ethanol 195

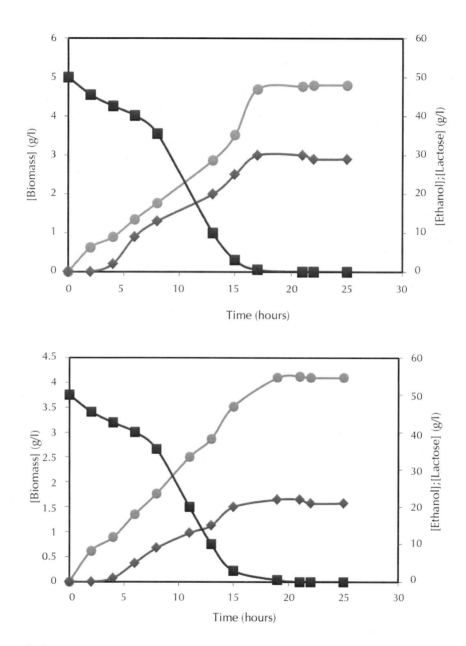

FIGURE 19.6 Cell (●), lactose (■) and ethanol (♦) concentration profiles for ordinary whey media with 50 g/l lactose with (a) aeration rate 0.4 vvm (b) aeration rate 0.8 vvm in airlift bioreactor.

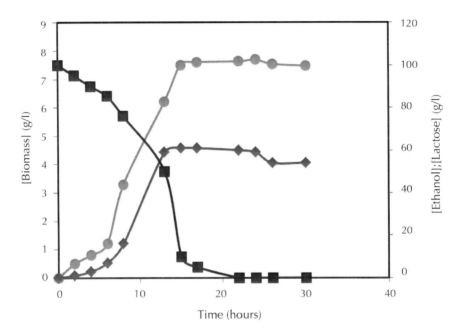

FIGURE 19.7 Cell (●), lactose (■) and ethanol (♦) concentration profiles for concentrated whey media with aeration rate 0.4 vvm and lactose 100 g/l.

KEYWORDS

- **Erlenmeyer flask**
- **Ethanol**
- **Lactose**
- **Peptides**
- **Pre-culture**

REFERENCES

Box, G., Hunter, W., and Hunter, S. (1978). Factorial Design at Two Levels. In: *Statistics for Experimenters, an Introduction to Design, Data Analysis and Model Building*. John Wiley and Sons, New York, pp. 306–644.

Domingues, L., Dantas, M. M., Lima, N., and Teixeira, J. A. (1999). Continuous ethanol fermentation of lactose by a recombinant flocculating *Saccharomyces cerevisiae* strain. *Biotechnol. Bioeng.* **64**, 692–697.

Domingues, L. L., Lima, N., and Teixeira, J. A. (2001). Alcohol production from cheese whey permeate using genetically modified flocculent yeast cells. *Biotechnol. Bioeng.* **72**, 507–514.

Dubois, M., Gilles, K. A., Hamilton, J. K., Rebers, P. A., and Smith, F. (1956). Colorimetric method for determination of sugars and related substances. *Anal. Chem.* **8**, 350–366.

Ferrari, M. D., Loperena, L., and Varela, H. (1994). Ethanol production from concentrated whey permeate using a fed-batch culture of *Kluyveromyces fragilis*. *Biotechnol. Lett.* **16**, 205–210.

Gawel, J. and Kosikowski, F. V. (1978). Improving alcohol fermentation in concentrated ultrafiltration permeates of cottage cheese whey. *J. Food Sci.* **43**, 1717–1719.

Janssens, J. H., Bernard, A., and Bailey, R. B. (1984). Ethanol from whey-continuous fermentation with cell recycle. *Biotechnol. Bioeng.* **26**, 1–5.

Kosikowski, F. V. (1979). Whey utilization and whey products. *J. Dairy Sci.* **62**, 1149–1160.

Moulin, G., Guillaume, M., and Galzy, P. (1980). Alcohol production by yeast in whey ultrafiltrate. *J. Biotechnology and Bioengineering* **22**, 1277–1281.

OECD-FAO Agricultural Outlook 2008–2017 Highlights. Paris: Organization for Economic Co-operation and Development—Food and Agriculture Organization of the United Nations, 2008 (available online b http://www.agri-outlook.org/dataoecd/54/15/40715381.pdf N last visited: 1st June 2009).

Ryu, Y. W., Jang, H. W., and Lee, H. S. (1991). Enhancement of ethanol tolerance of lactose assimilating yeast strain by protoplast fusion. *J. Microbiol. Biotechnol.* **1**, 151–156.

Silveira, W. B., Passos, F. J. V., Mantovani, H. C., and Passos, F. M. L. (2005). Ethanol production from cheese whey permeate by *Kluyveromyces marxianus* UFV-3: A flux analysis of oxido-reductive metabolism as a function of lactose concentration and oxygen levels. *Enzyme Microb. Technol.* **36**, 930–936.

Siso, M. I. G. (1996). The biotechnological utilization of cheese whey: A review. *BioresTechnol* **57**, 1–11.

Siso, M. I. G., Ramil, E., Cerdan, M. E., and Freire Picos, M. A. (1996). Respirofermentative metabolism in *Kluyveromyces lactis*: Ethanol production and the Crabtree effect. *Enzyme Microbiol Technol* **18**, 585–591.

Smithers, G. W. (2008). Whey and whey proteins-from 'gutter-to-gold'. *Int. Dairy J.* **18**, 695–704.

Terrel, S. L., Bernard, A., and Bailey, R. B. (1984). Ethanol from whey: Continuous fermentation with a catabolite repression-resistant *Saccharomyces cerevisiae* mutant. *Appl. Environ. Microbiol.* **48**, 577–580.

Index

A

Activity coefficient, 96–97, 102, 103–104
Airlift bioreactor, 187
Aqueous solutions, 61, 173
Area fraction, 68
Artificial neural networks (ANNs), 11, 126
  activity coefficient, 92
  advantage, 93
  application for, 166
  architecture of, 93
  Bayesian regularization, 167
  biological neurons, 166
  compressibility factor, 166
  equations of state (EOS), 91
  equilibrium temperature, 92
  experimental equilibrium data, 169
  FNNs, 166
  group method of data handling (GMDH)
    algorithm, 50–53
    LLE prediction, 53–58
  human brain, 166
  hydrofluoroethers (HFEs), 166
  independent and dependent variables, 92
  iterative method, 91
  Levenberg-Marquardt algorithm, 167
  liquid and vapor phase, 92
  LLE measurements, 49
  mean absolute error (MAE), 170
  mean square error (MSE), 170
  model, 167
  mole fractions, comparison of, 167–168
  multi-layer perceptron (MLP)
    input and output parameters, 93–94
    single hidden layer, 95
    transformation functions, 94
  PNNs and RBF, 166
  prediction, 169–170
    equilibrium data, 168
    method, 50
  steps in, 166
  thermodynamic
    models, 49
    properties, 92, 166
  thermodynamics, 165
  vapor-liquid equilibrium (VLE), 166
    estimation, 91

B

Bayesian regularization, 167
Binary systems, densities and refractive indices
  chemical
    Abbe refractometer, 44
    Julabo circulator, 44
  liquid mixtures, 43
  measurements, 43–44
  mole fraction, 46
  ternary mixtures, 44
  thermodynamic properties, 43
Binoda (solubility) curves, 148
Biochemical oxygen demand (BOD), 185
Boiling temperature, 97–98
Bulk liquid membrane
  aqueous phases, 77
  organic membrane phase, 76

C

Calibrated digital thermometer, 102
Calibration coefficients, 137
Central composite design (CCD), 188
Cheese whey, 185
Cheese whey permeate (CCWP), 186
Copper-constantan thermocouple, 103

D

Densities and excess molar volumes
  atmospheric pressure, 7
  binary systems, 8–9
  experimental values, 8
  polynomial equations, 7
  thermodynamic interactions, 7
Diphenylcarbazide method, 80

E

Equilibrium models, 28
Ethanol, laboratory production
   aeration rate on, 190–196
   Candida pseudotropicalis, 186
   CCD, 188
   CCWP, 186
   cellular osmotic pressure, 190
   cheese whey, 185
   COD and BOD, 185
   coefficient estimates, in second-order model, 189
   culture media and batch cultures
     airlift bioreactor, 187
     fermentation, 186
     microorganism, 186
     preculture, 186
   factors levels, 189
   fermentation process, 186, 188
   interruption coefficient of, 188
   Kluyveromyces fragilis, 186
   lactose concentration, 188
   productivity of, 186
   quadratic effect, coefficient of, 188
   response surface curves, 189
   second degree polynomials, 188
   sugar concentrations, 188
   surface methodology, 189
   whey media fermentation, 192
   yeast cells concentration, 188
Etherification reaction, 1–2
Extraction processes, 61

F

Feedforward neural networks (FNNs), 166
Fermentation technology, 135
First-order perturbation theory, 116–117
Fluid phase equilibrium. See also Molecular thermodynamics process control in
   area fractions, 103
   calibrated digital thermometer, 102
   cell, 102
   copper-constantan thermocouple, 103
   cumene, 102
   organic-rich phase, 103
   physical properties of, 103
   purity of, 103
   refractive indexes, 103
   solubility of water in cumene, 106
   thermal conductivity detector (TCD), 103

G

Gibbs free energy, 137
GPSA Engineering Data Book, 158
Group method of data handling (GMDH), 12
   distribution coefficient, 131, 132
   experimental data points, 128
   LLE data, 129
   mathematical model of, 127
   mole fraction values, 128
   multivariate analysis method, 127
   neural network type
     nodes, 127
     structure of, 128, 129
     testing and training data, 127–128
   prediction of experimental data, 129–131
   RMSD, 131–132
   Volterra Series, 127

H

Hydrofluoroethers (HFEs), 166
HYSYS and UNIQUAC models
   apparatus and procedure, 174
   experimental and predicted LLE data, 175–177
   extracted phase, 180
   materials, 174
   Othmer and Tobias correlation, 180
   root mean square deviation (RMSD), 179
   section and area fraction, 175
   separation factor, 180
   simulation and separation program
     acetic acid purification, 181
     distillation column, 181
     flow diagram, 181
     materials and stream compositions, 181–182
     operation conditions, 181

K

Kolmogorov-Gabor polynomial, 50. See also Volterra Series
Konik gas chromatography (GC), 103

L

Least-squares technique, 53
Levenberg-Marquardt algorithm, 167
Liquid-liquid equilibrium (LLE), 11, 14, 18, 28, 39, 49, 53, 55, 66–71, 102, 108, 126, 128, 137, 147–151, 165, 167
  ANNs, 126
  GMDH, 126, 127–132
    algorithm, 12–13
    formal definition, 12
    inputs and output variables, 12–13
    Kolmogorov-Gabor polynomial, 13
    least squares technique, 14
    regression techniques, 13
  prediction
    aqueous and organic phases, 15
    calculated points, 20–21
    developed structure, 18
    experimental compositions, 14
    experimental points, 20–21
    GMDH models, 24
    GMDH-type neural network, 14
    optimal structures, 16
    polynomial equations, 17
    root-mean-square deviation (rmsd), 22
    separation factors, 22
    statistical values, 19
    ternary system, 22
    UNIFAC model, 14, 22
  systems, 66
  thermodynamic models, 126
  VLE, 126
Liquid membrane separation process
  bulk liquid membrane
    aqueous phases, 77
    organic membrane phase, 76
  chromium transportation, 79
  diphenylcarbazide method, 80
  error analysis, 78
  experiment, 77
  mixing effect, 85–87
  model experiment
    feed phase, 78
    stripping phase, 78
  molarity effect, 87–88
  parameters, 78
  phase for effect of percentage, 80–83
  pH effect, 85
  volumetric percentage, 85
LLE. See Liquid-liquid equilibrium (LLE)

M

Marek and Standart (M-S) method, 4
Mass fractions, 53–54, 57
Maxwell's distribution, 121
Mean absolute deviation (MAD), 12
Mean absolute error (MAE), 170
Mean square error (MSE), 170
Mixer rotational rate, 76
Molecular thermodynamics process control in
  compressibility factors, 116
  cubic equations of state, 116
  equation of state, 116–120
  first-order perturbation theory, 116–117
  gas pressure gradient, 116
  supercritical and subcritical, 116
  theoretical model for covolume term
    coefficient of polynomial, 121
    collision energy, 120
    decay function, 122
    equations of state, 121–122
    Kihara's model, 120
    Maxwell's distribution, 121
    potential function, 120
    translational energy, 121

N

Neural networks (NNs), 11–12
Non-random two-liquid (NRTL) model, 11
  cell temperature, 62
  chromatographic analysis, 62
  estimation procedures, 62
  gravimetrical analysis, 62
  mole fractions, 63
  quaternary system, 61
  volatile compounds, 62

O

Optimization of process
  advantages of, 136
  amine's solvation power, 136
  calcium hydroxide precipitation method, 136
  calibrated digital thermometer, 136
  calibration coefficients, 137
  experimental apparatus of, 136
  fermentation technology, 135
  gas chromatography, 136
  Gibbs free energy, 137
  LLE measurement
    correlation of, 139–141
    experimental and predication data, 137–139
    extraction of propionic acid, 137
    indexes components and phases, 142
    mole fraction, 137
    Othmer and Tobias correlation, 143
    rmsd, 142
    separation factor, 142
    structural parameters, 141
    tie-lines, 142
    UNIQUAC model parameters, 141
  mole fractions, 137
  petroleum feedstocks, 135
  phase separation, 136
  propionic acid, 135–136
  reactive extraction, 136
  simulation and separation program
    materials and stream compositions, 144
    Othmer-Tobias equation constants, 144
    propionic acid extraction process, 143–144
    purification process, 143–144
  thermal conductivity detector (TCD), 136–137
  UNIQUAC model, 137
  whey lactose, 136
Organic solvents, 66, 173
Othmer and Tobias correlation, 111, 143, 153, 180
Oxygenated compounds, 29

P

Perturbation theory, 117
Petroleum feedstocks, 135
Phase compositions, 174
Probabilistic neural networks (PNNs), 166

Q

Quaternary system, 50

R

Radial basis function (RBF), 166
Redlich-Kister equation, 43–44
Refractive indexes, 103
Root mean square deviation (rmsd), 109–110, 131–132

S

Salting-out and -in, 61
Segment fraction, 68
Shimadzu C-R2AX integrator, 67, 103
Sigmoid function, 156
Sour gas
  artificial neural networks (ANN)
    comparison with Wichert and BM methods, 159–160
    input-output properties of, 157
    model predictions, water content, 160–161
    multi-inputs one-output neuron, 156
    multilayer perceptron (MLP), 157
    neuron, 156
    non-linear activation function, 156
    number of hidden nodes, 158
    parameters, 158
    sigmoid function, 156
    variables of, 158, 159
  gas mixture water content, 161
  GPSA Engineering Data Book, 158
  sweet gas contribution, 156
  water condensation, 155
  water content of, 155–156
  Wichert and BM methods, 159
  Wichert correlation, 159
Square-well (SW) potential, 118
SRK and PR equations, 118
State equation

Index

compressibility factor, 117
integration constant, 120
perturbation theory, 117
potential energy, 118
repulsive effects, 117
square-well (SW) potential, 118
SRK and PR equations, 118
thermodynamic properties, 118
truncated virial series, 119
Stripping phase molarity, 76
Supercritical and subcritical, 116

T

Ternary systems, 11, 61, 173
Thermal conductivity detector (TCD), 67, 103, 149
Thermodynamic model
   molar volumes, 3
   regression, 3
   VLE computation, 2
Training and test data, 51, 53

U

UNIFAC models
   activity coefficient, 102, 103–104
   area fraction, 104
   functional group, 102
   group parameters, 105
   interaction parameters, 104, 105
      activity coefficients, 149
      binoda (solubility) curves, 148
      combinatorial and residual parts, 149
      correlated tie lines, 150–151
      distribution coefficients, 152
      experimental tie line data, 150–151
      fluid-phase equilibrium, 147
      gas chromatography, 148
      interaction parameters, 150
      mole fractions, 149
      Othmer-Tobias equation constants, 111, 152–153
      phase separation, 148
      separation factor, 152
      solvent, selectivity and strength, 151
      Star integrator, 149
      thermal conductivity detector, 149
      tie-line data, 149

lattice coordination number, 104
LLE data, 104
mole fractions of, 103–104
multi-component systems, 102
predicted tie lines, 106–108
pure component constant, 104
rmsd, 109–110, 153
segment fraction, 104
separation factor, 110
solvent selectivity and strength, 106
tie lines of system, 108–109
van der Waals group volume, 105
Universal quasi-chemical (UNIQUAC)
   method, 11
   models, 57, 137
      adjustable parameter, 68
      area fraction, 68
      chemicals, 66
      equilibrium model, 68
      gas chromatography, 67
      interaction parameters, 66, 68, 72
      procedure, 66–67
      rmsd, 70
      segment fraction, 68
      structural parameters, 69–70

V

van der Waals group volume, 105
Vapor-liquid and liquid-liquid equilibrium
   activity coefficients, 32
   antoine coefficients, 35
   apparatus and procedure, 29–30
   area fraction, 38
   aromatic compounds, 29
   binary interaction parameters, 36
   boiling temperature diagram, 33–35
   combinatorial parte, 38
   liquid-liquid phase diagrams, 38
   materials, 29
   paraffin, 29
   precise LLE data, 28–29
   relative error, 40
   segment fraction, 38
   separation factor, 40
   ternary systems, 38
   UNIQUAC model, 37, 39
      structural parameters, 36

Vapor-liquid equilibrium (VLE), 2, 4, 11, 27–31, 40, 49, 91–92, 94, 98, 126, 165
investigations, 91
Volterra Series, 50, 127

W

Water condensation, 155
Water + propanoic acid+ 1-octanol system. See Optimization of process
Whey media fermentation, 192
Wichert and BM methods, 159
Wichert correlation, 159
Wilson parameters, 4